Contents

Appendix

Introduction to Chemistry for Biology Students

Sixth Edition

George I. Sackheim
Associate Professor Emeritus, Chemistry
University of Illinois at Chicago

formerly
Coordinator of Biological and Physical Sciences
Michael Reese Hospital and Medical Center, Chicago

Science Instructor
St. Francis Hospital School of Nursing
Evanston, Illinois

An imprint of Addison Wesley Longman, Inc.

Menlo Park, California • Reading, Massachusetts • New York • Harlow, England
Don Mills, Ontario • Sydney • Mexico City • Madrid • Amsterdam

Publisher: *Robin J. Heyden*
Acquisitions Editor: *Ben Roberts*
Production Editor: *Lisa Weber*
Composition and Film Supervisor: *Vivian McDougal*
Manufacturing Supervisor: *Holly Bisso*
Market Development Manager: *Andy Fisher*
Director of Marketing: *Stacy Treco*
Channel Marketing Manager: *Gay Meixel*
Copy Editor: *Anita Wagner*
Proofreader: *Rosemary Sheffield*
Text Designer: *Patrick Devine*
Illustrators: *Dovetail Publishing Services and Val Felts*
Compositor: *Dovetail Publishing Services*
Cover Designer: *Yvo Riezebos*
Cover Photograph: *© 1998 PhotoDisc, Inc.*

Credits: Pages 185 and 278: Figures a and b from W. M. Becker and D. W. Deamer, *The World of the Cell*, 3rd ed. (Menlo Park, CA: Benjamin/Cummings, 1996). © 1996 Benjamin/Cummings Publishing Company; Figures d and e © Irving Geis. Appendix figures A, B, C, and D adapted from N. A. Campbell, L. G. Mitchell, and J. B. Reece, *Biology: Concepts and Connections,* 2nd ed. (Menlo Park, CA: Benjamin/Cummings, 1997). © 1997 Benjamin/Cummings Publishing Company.

Library of Congress Cataloging-in-Publication Data
Sackheim, George I.
     Introduction to chemistry for biology students / George I.
  Sackheim. — 6th ed.
       p.   cm.
     Includes index.
     ISBN 0-8053-3097-6
     1. Chemistry—Programmed instruction.   I. Title.
  QD33.S134   1998
  540'.7'7—dc21   98-30690
                                            CIP

1 2 3 4 5 6 7 8 9 10—CRW—-02 01 00 99 98

The Benjamin/Cummings Publishing Company, Inc.
2725 Sand Hill Road
Menlo Park, CA 94025

To the Student

Introduction to Chemistry for Biology Students, Sixth Edition, is not an ordinary book. It has been programmed to help you review the basic facts, concepts, and terminology of chemistry that are essential to an understanding of biological phenomena. Today's biology courses place increasing emphasis on the chemical processes that underlie the critical biological functions. This book will help you to understand those processes.

The topics involved are among the most critical and exciting that science will explore in the years ahead. What are the basic chemical processes underlying biological phenomena? What are the essential differences between living and nonliving matter? What are the conditions under which molecules organize into living matter? Can these conditions be duplicated experimentally?

If you have already had a course in chemistry, this programmed book can serve as an effective review of the fundamental concepts. If you have had no previous chemistry, the program will give you the background you need to gain a clear understanding of the biological processes you will be studying.

Do not try to skip around in this book. Start at the beginning and proceed through it at the pace that is most comfortable for you. Under no circumstances should you rush through. Whether it takes you two hours or six hours to complete is of little importance. Your primary objective should be to master the material no matter how long that may take.

The material covered in this book will help you most if you complete it during the first two weeks of your biology course. Having done that, you will be ready to handle the chemical aspects of biology as they come up.

When you have completed the program, you may want to repeat certain material. To simplify this process, use the book's Contents to help you locate specific topics.

If you follow the directions and complete this program, you will learn to:

- recognize elements present in various compounds

- know what is meant by pH and by ionization

- recognize acids, bases, and salts

- discriminate between electrolytes and nonelectrolytes

- compare osmosis and diffusion

- distinguish between passive and active transport

- understand osmotic pressure

- understand transmission of nerve impulses, including depolarization and repolarization

- understand how neurotransmitters work

- know how DNA replicates

- know how mRNA is formed and how it regulates protein synthesis

- understand oxidation and reduction

- know what isotopes are

- recognize various organic functional groups

- differentiate among carbohydrates, fats, and proteins

- understand how enzymes function

- recognize nucleic acids

- understand biologic oxidation, including glycolysis, the Krebs cycle, and the electron transport chain

- understand reactions involved in photosynthesis, including light reactions and the Calvin cycle

- follow the flow of oxygen from the lungs to cells and of carbon dioxide from cells to the lungs on gas partial pressure gradients

How to Use This Book

This type of instructional book may be new to you. Its subject matter has been presented as a series of numbered problems. Each builds on information you have learned in the preceding problems. For that reason, it is important not to skip around. The sequence of the problems is important because it is programmed to help you learn more efficiently.

Respond to Every Problem

Some problems present new information; others review material presented earlier. Every problem presents a learning task that requires some response from you.

You may be asked to make any of the following types of responses:

- write an answer in a blank space

- label a diagram

- draw a simple diagram

- select the correct answer from among several alternatives

- write a sentence in answer to a question

When you have written or marked your answer, you will want to find out whether you are correct. Programmed instruction provides you with important feedback by giving you easy access to the answers, which are located in shaded areas on the outer side of each page. *Do not look at the correct answer until after you have marked your own answer.* If you look before answering, you will only impair your own learning process.

Use an Answer Mask

Bound into the back of the book is a sheet of heavy paper that is perforated. Tear off the outer portion of the sheet for use as an answer mask. Here is what to do:

1. As you start working on the first page, a right-hand page, cover the shaded answer column with the answer mask *before* looking at the problems. When you turn the page, shift your answer mask to cover the answer column on the left-hand page before reading the problems on that page. *Be sure you have covered the answers before you read anything.*

2. Each problem number is centered on the page with a shaded rectangle behind it. Read the problem carefully, then record your answer. Make sure you either write each answer or do whatever the directions say. Do not simply think of the answer and then go on. Actually writing or marking your answer reinforces your learning.

3. Now move the answer mask aside to reveal the answer, which you will find aligned with the problem number. (If the main part of the problem runs onto the next page, you will find the answer at the top of that page.)

4. If your answer was correct, move on to the next problem.

5. If your answer was incorrect, reread the problem (if necessary, reread several of the preceding problems) until you understand your error and know why the given answer is correct. Then go on.

When you have completely worked your way through this book, you should have the knowledge of chemistry you need to succeed in your biology courses.

PART I

Inorganic
Chemistry

Atomic Structure 1

ELEMENTARY PARTICLES

Atoms are made up of several components. Collectively these components are called the *elementary particles*. We will be discussing the three major elementary particles: *protons, neutrons*, and *electrons*.

Here is a diagram of an atom:

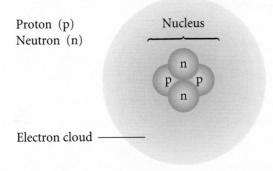

Proton (p)
Neutron (n)

Nucleus

Electron cloud

1

The protons (p) and the neutrons (n) are packed together in an inner core called the _____. The outer part of the atom, which contains electrons, is called the _____.

a negative electrical
charge, because the
electron cloud consists of
electrons

2

The electron cloud has a negative electrical charge. What type of charge would you expect the electron to have? _____

_____

a. repel
b. attract
c. repel
d. attract

3

The electron has a *negative electrical charge* and is symbolized by e⁻. Remember that *like* electrical charges repel each other, and *unlike* charges attract.

Indicate whether the following pairs of charges would attract or repel each other.

a. ⊕ ⊕ _____   b. ⊖ ⊕ _____

c. ⊖ ⊖ _____   d. ⊕ ⊖ _____

positive

4

The nucleus attracts the negatively charged electrons. Therefore, the overall charge of the nucleus must be _____ (negative/positive).

proton

5

The neutron was named for its electrical characteristics. It has no electrical charge; it is neutral. This means that the positive charge of the nucleus must be due to the *second type of particle* it contains. This second type of particle is the _____.

6

So far, then, we have this picture of atomic structure:

a.   An atom consists of an inner part, or _____, that is made
     up of _____ and _____.

b.   The electron has what type of charge? _____

c.   The proton has what type of charge? _____

d.   The neutron has a charge of _____.

a. nucleus; protons;
    neutrons
b. negative
c. positive
d. zero (0)

7

The charge on the electron balances the charge on the proton. If the
electron has a charge of −1, then the proton would have a charge of
_____ ($-1, +1, \pm1$).

+1

8

An atom with one proton in its nucleus and one electron outside
that nucleus would therefore have an overall charge of _____
($+1, -1, 0$).

0

9

Atoms are electrically neutral. This means that an atom will contain:
(check one)

__ more protons than electrons
__ more electrons than protons
__ an equal number of protons and electrons

an equal number of
protons and electrons

12

An atom with 12 protons in the nucleus would have how many electrons outside the nucleus? _____

a. proton
b. electron

The atom with the simplest atomic structure is hydrogen. (For simplicity we shall merely indicate the electron(s) outside the nucleus, and omit the electron cloud.)

$$p \quad e^-$$

Hydrogen

a. The nucleus of the hydrogen atom consists of one _____.

b. The outer part of the atom, the electron cloud, contains one _____.

2; 2; 2

The helium atom is a little more complicated.

$$e^- \quad p \;\; p \quad e^-$$

Helium

It contains: (how many?)

_____ neutrons
_____ protons
_____ electrons

ATOMIC NUMBER

There are over 100 known elements. Each element has two numbers associated with it—numbers that give certain facts about the structure of its atoms.

The first number is the atomic number. This is the number of protons in the nucleus of the atom.

13

Hydrogen, the simplest atom, contains only one proton, so the atomic number of hydrogen is _____.

1

14

Uranium is the most complicated of the elements that occur naturally. A uranium atom contains 92 protons, 146 neutrons, and 92 electrons. The atomic number of uranium is _____.

92

15

An atom of magnesium, atomic number 12, must have a nucleus containing _____ protons.

If the nucleus contains 12 protons, there must be how many electrons? _____

12; 12

16

Therefore, the atomic number of an element indicates the number of _____ in the nucleus of the atom, and also the number of _____ outside the nucleus.

protons; electrons

MASS NUMBER

The second number associated with each atom is the *mass number*. The mass number expresses the sum of the masses of the particles in the atom.

A proton has a mass of 1 atomic mass unit. An electron is considered to have zero mass, or a mass of 0.

1

17

A hydrogen atom has a mass of _____. (If you don't know, see problem 11.)

a. 2
b. 0
c. 2
d. 1

18

The helium atom has a mass number of 4.

a. The 2 protons in the helium atom have a total of how many atomic mass units? _____

b. The 2 electrons in the helium atom have a total of how many atomic mass units? _____

c. Therefore, for the helium atom to have a mass number of 4, the 2 neutrons must contain how many atomic mass units? _____

d. If 2 neutrons have a total of 2 atomic mass units, a neutron must have an atomic mass of _____.

in the nucleus

19

Because the electrons, which have practically no mass, are located outside the nucleus, the entire mass of the atom can be considered to be located:

__ in its electron cloud   __ in the nucleus

20

The *atomic number* indicates the number of protons (each with atomic mass 1) inside the nucleus of an atom. The *mass number* indicates the number of protons and neutrons (each with atomic mass 1) in the nucleus. Therefore, the number of neutrons can be determined by *subtracting* the atomic number from the mass number.

The sodium atom has a mass number of 23 and an atomic number of 11. The number of neutrons in the nucleus of the sodium atom is

_____.

12

21

The carbon atom has an atomic number of 6 and a mass number of 12. The carbon atom contains: (how many?)

_____ protons in its nucleus
_____ neutrons in its nucleus
_____ electrons outside its nucleus

6; 6; 6

22

The element phosphorus has an atomic number of 15 and a mass number of 31. Indicate on the blank lines in the diagram the number of protons, neutrons, and electrons.

Phosphorus

15p
16n   15e⁻

12e⁻

$$23$$

Diagram the structure of the magnesium atom, atomic number 12 and mass number 24.

Magnesium

92; 238

$$24$$

The uranium atom has the atomic structure shown here.

92e⁻

Uranium

Therefore, the uranium atom has an atomic number of _____ and a mass number of _____.

ISOTOPES

$$25$$

a.   Draw the structure of a carbon atom, atomic number 6 and mass number 12.

a.

6p
6n   6e⁻

Carbon-12

b. Draw the structure of a carbon atom, atomic number 6 and mass number 13.

b.

Carbon-13

a. **the same**
b. **different**

26

Here are the structures you drew for the two carbon atoms:

Carbon-12   Carbon-13

a. These atoms have _____ (the same/different) atomic number(s).

b. These atoms have _____ (the same/different) mass number(s).

Such atoms are called *isotopes*.

27

Isotopes, then, may be defined as atoms that have:

__ the same atomic number and the same mass number
__ different atomic numbers
__ different mass numbers and the same atomic number

**different mass numbers
and the same atomic
number**

Chlorine-35

Chlorine-37

Hydrogen-1

Hydrogen-2

Hydrogen-3

a. **same**
b. **different**

28

Draw the two isotopes of chlorine, atomic number 17 and mass numbers 35 and 37.

a. chlorine-35   b.   chlorine-37

 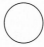

29

Draw the three isotopes of hydrogen, atomic number 1 and mass numbers 1, 2, and 3.

a. hydrogen-1   b.   hydrogen-2   c.   hydrogen-3

30

Radioisotopes, isotopes that give off radiation, are frequently used in medical applications. I-131 (radioactive iodine, also written as ^{131}I) is used in the diagnosis and treatment of thyroid conditions. How does I-131, radioactive iodine, compare with I-127, nonradioactive iodine, in

a.   atomic number _____

b.   mass number _____

ELECTRON ENERGY LEVELS

The electrons are located outside of the nucleus of the atom. These electrons make up the electron cloud, which may be subdivided into different energy levels. The first energy level is nearest the nucleus; then comes the second energy level, the third energy level, and so on.

Each energy level can hold a certain maximum number of electrons. This maximum number may be determined by using the formula $X = 2n^2$ (X is the maximum number of electrons in energy level number n).

31

2; 2

Using the formula $X = 2n^2$, if $n = 1$, then $X =$ _____. The energy level indicated by $n = 1$ is the first energy level. Therefore, the first energy level can hold a maximum of _____ electrons.

32

a. yes
b. no

a.   Can the first energy level hold fewer than two electrons? _____

b.   Can the first energy level hold more than two electrons? _____

33

8

For the second energy level, where $n = 2$, the maximum number of electrons is _____.

34

a. 18
b. 32

a.   The maximum number of electrons in the third energy level is

_____.

b.   The maximum number of electrons in the fourth energy level is

_____.

2e⁻ in the first energy level; 8e⁻ in the second energy level; 18e⁻ in the third energy level; 32e⁻ in the fourth energy level

8

35

Label the maximum number of electrons possible in each energy level in the diagram.

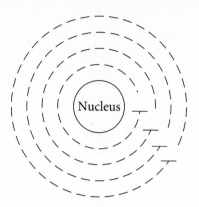

36

The following rules must be observed when considering the placement of electrons in the various energy levels. The first energy level must be filled with 2 electrons before electrons can go into the second energy level.

The second energy level must be filled with _____ electrons before electrons can go into the third energy level.

For elements having more than three energy levels, the sequence of filling those levels is complex and can be found in a general chemistry text.

37

On the diagram, show the structure of the hydrogen atom, atomic number 1 and mass number 1.

Hydrogen atom

a. The number of protons in the hydrogen atom is _____.

b. The number of neutrons is _____.

c. The number of electrons is _____.

d. The 1 electron in the hydrogen atom must go into which energy level—1st, 2nd, or 3rd? _____

38

In the space provided, draw the structure of the helium atom, atomic number 2 and mass number 4.

Helium atom

Hydrogen atom

a. 1
b. 0
c. 1
d. 1st

Helium atom

a. 3
b. 4
c. 3

2

Lithium atom

39

Now let's look at the structure for the lithium atom, atomic number 3 and mass number 7.

a.   The number of protons in the lithium atom is _____.

b.   The number of neutrons in the lithium atom is _____.

c.   The number of electrons in the lithium atom is _____.

40

There are 3 electrons in the lithium atom. How many energy levels will the lithium atom have? _____ (If you aren't sure, check problem 36.)

41

Complete the diagram of the lithium atom.

Lithium atom

42

The carbon atom has atomic number 6 and mass number 12.

a. The carbon atom contains _____ protons.

b. The carbon atom contains _____ neutrons.

c. The carbon atom contains _____ electrons.

d. Complete the structure of the carbon atom.

Carbon atom

a. 6
b. 6
c. 6

Carbon atom

43

a. In the element sodium, atomic number 11, there are how many electrons? _____

b. How many energy levels will the sodium atom have? _____

c. How many electrons will each energy level of the sodium atom have?

   1st energy level __   2nd energy level __   3rd energy level __

d. Diagram the complete structure of the sodium atom, atomic number 11 and mass number 23.

Sodium atom

a. 11
b. 3
c. 2; 8; 1

Sodium atom

Silicon atom

Diagram the structure of the silicon atom, atomic number 14 and mass number 28.

Chemical Symbols 2

Elements are the building blocks of the universe. They cannot be broken down into simpler substances by ordinary chemical means. Each of the over 100 elements has its own name and its own properties. The symbol for an element is usually an abbreviation for its name. Frequently the abbreviation is merely the first letter of that name. The following table lists some of these elements.

Element	Symbol
Hydrogen	H
Carbon	C
Oxygen	O
Nitrogen	N
Phosphorus	P
Sulfur	S

45

What elements are present in water, H_2O? _____

hydrogen and oxygen

46

What elements are present in ammonia, NH_3? _____

nitrogen and hydrogen

carbon, hydrogen, and
oxygen

47

What elements are present in glucose, $C_6H_{12}O_6$? _____

carbon, hydrogen,
oxygen, nitrogen,
sulfur

48

Biotin, one of the B-vitamins, $C_{10}H_{16}O_3N_2S$, contains which elements?
_____

carbon, hydrogen,
oxygen, nitrogen,
phosphorus

49

Adenosine triphosphate, ATP, the body's primary energy compound, has the formula $C_{11}H_{18}O_{13}N_5P_3$. What elements does it contain? _____

Ca; Br; Si; Ba

50

When the names of more than one element begin with the same letter, frequently the second letter of the name is added to the symbol. Note that only the first letter of the symbol is capitalized.

Here are some of the elements whose symbols are the first two letters of their names. Write the symbols for each.

Element	Symbol
Calcium	_____
Bromine	_____
Silicon	_____
Barium	_____

calcium and bromine

51

$CaBr_2$ contains which two elements? _____

52

BaO_2 contains which two elements? _____

barium and oxygen

53

CaC_2O_4 contains which three elements? _____

calcium, carbon, and oxygen

There are several elements whose symbols are not derived from the first letter or first two letters of their English names. Some of those elements are listed in the following table.

Element	Symbol
Chlorine	Cl
Sodium	Na (from the Latin *natrium*)
Magnesium	Mg
Potassium	K (from the Latin *kalium*)
Zinc	Zn
Iron	Fe (from the Latin *ferrum*)

54

Write the name of the element represented by each of the following symbols:

H _____	Na _____	Si _____
N _____	K _____	Br _____
O _____	C _____	Zn _____
Ca _____	S _____	Fe _____

Hydrogen; Nitrogen; Oxygen; Calcium; Sodium; Potassium; Carbon; Sulfur; Silicon; Bromine; Zinc; Iron

a. potassium and bromine
b. carbon and chlorine
c. sodium, oxygen, and hydrogen
d. magnesium and chlorine
e. sodium, nitrogen, and oxygen
f. potassium, iron, carbon, and nitrogen

carbon, hydrogen, oxygen, nitrogen, and magnesium

55

What elements are present in each of the following?

a. KBr _____

b. CCl_4 _____

c. NaOH _____

d. $MgCl_2$ _____

e. $NaNO_3$ _____

f. $K_3FeC_6N_6$ _____

56

Chlorophyll has the formula $C_{55}H_{68}O_5N_4Mg$. It contains which elements? _____

Atoms and Molecules 3

57

One atom of hydrogen is written as H. The number 1 is understood. Two atoms of hydrogen are written as 2H. Three atoms of carbon are written as _____.

3C

58

Atoms combine to form *molecules*. A molecule contains two or more atoms of the same or different elements. The molecule O_2 contains how many atoms of oxygen? _____

2

59

Each molecule of glucose, $C_6H_{12}O_6$, contains:

_____ atoms of carbon
_____ atoms of hydrogen
_____ atoms of oxygen

6; 12; 6

60

A water molecule is composed of two hydrogen atoms and one oxygen atom. Write the formula for a water molecule. _____

H_2O

$C_{12}H_{22}O_{11}$

61

A molecule of sucrose contains 12 atoms of carbon, 22 atoms of hydrogen, and 11 atoms of oxygen. The formula for sucrose is _____.

$C_{63}H_{90}O_{14}N_{14}PCo$

62

One molecule of vitamin B_{12} contains:

63 atoms of carbon
90 atoms of hydrogen
14 atoms of oxygen
14 atoms of nitrogen
 1 atom of phosphorus
 1 atom of cobalt (Co)

The formula for vitamin B_{12} is _____.

a. 6
b. $8H_2O$

63

When you want to represent more than one molecule of a compound, the number of molecules precedes the symbols.

a.   Thus, $6CO_2$ indicates how many molecules of carbon dioxide? _____

b.   How would eight molecules of water be represented? _____

molecule; hydrogen; 2

64

The formula H_2 indicates one _____ of _____, composed of how many atoms? _____

65

The formula $2H_2$ indicates _____ of hydrogen.

66

What do the following formulas represent?

3O _____ $4N_2$ _____ $7H_2O$ _____

two molecules

3 atoms of oxygen;
4 molecules of nitrogen;
7 molecules of water

Ionization 4

IONS

67

Atoms contain equal numbers of protons and electrons. Thus, atoms are:

__ electropositive   __ electronegative   __ electrically neutral

electrically neutral

68

The electron is a _____ (negatively/positively) charged particle.
So, if an atom loses an electron, it will then have an overall _____
(negative/positive) charge.

negatively; positive

69

If an atom were to *gain* an electron, it would then have an overall
_____ charge.

negative

70

An *ion* is an atom that has acquired an electrical charge by either losing
or gaining an _____.

electron

positively

71

Note that electrons, which are located outside the nucleus, are lost or gained in the formation of ions. The nucleus, which contains protons and neutrons, does not take part in the formation of ions.

If a hydrogen atom loses its electron, it becomes a hydrogen ion. Because it loses an electron, it is a _____ (negatively/positively) charged ion.

a. ion
b. negatively
c. positively

72

a.   Any atom that has lost or gained an electron is an _____.

b.   If an atom gains an electron, it becomes a _____ charged ion.

c.   If an atom loses an electron, it becomes a _____ charged ion.

valence shell

73

Some elements reach their most stable state when they have 2 electrons in their outer (their only) energy level, which is called the *valence shell*.

Helium atom

This atom of helium, for example, contains 2 electrons, which fill the first energy level. So, the first energy level in helium is called the

_____.

74

Helium (He) has a completed valence shell, one that can hold no more electrons. Thus, helium is a very _____ (stable/unstable) atom. Such elements are usually unreactive.

75

Most atoms reach their most stable state when they have 8 electrons in their outer energy level. This is called the octet rule.

Neon atom

This atom of neon (Ne), for example, contains 10 electrons. Two of the electrons fill the first energy level, and the remaining 8 electrons are in the second energy level. So the second energy level of neon would be called its _____.

stable

valence shell

a. valence shell; third; 1
b. Cl

76

Look at this atom of chlorine.

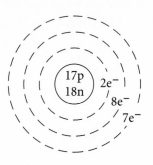

Chlorine atom

a.   In chlorine, the outer energy level, or _____, is the
     _____ energy level. It contains 7 electrons. Thus, to attain
     a stable state, the chlorine atom needs to gain how many electrons?
     _____

b.   What is the symbol for chlorine? _____

a. ion
b. negative

77

Suppose that an atom of chlorine gains the 1 electron it needs to become
stable.

a.   Because it has gained an electron, it is now a charged atom, called
     a(n) _____.

b.   Would it have a positive or negative charge? _____

Cl^-

78

Which of the following symbols stands for the chloride ion (a chlorine
atom that has gained 1 electron)?

__ Cl   __ Cl^+ __ Cl^-

79

If a chlorine atom is to gain an electron, some other atom must lose the electron.

a.   That atom could be sodium, the symbol for which is _____.

b.   The sodium atom has how many electrons in its valence shell?

_____

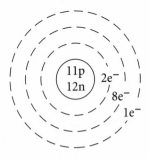

Sodium atom

c.   To become stable it will have to _____ (gain/lose) one electron.

80

When the sodium atom loses 1 electron, it becomes a _____ (positively/negatively) charged _____.

81

Which symbol should we use for the sodium ion?

__ Na   __ Na$^+$ __ Na$^-$

a. Na
b. 1
c. lose

positively; ion

Na$^+$

a. +1
b. −1
c. +2

82

Positive and negative charges are indicated by superscript plus (positive) or minus (negative) signs, respectively, with the number 1 being understood. Charges greater than 1 are written as superscripts, with the number preceding the plus or minus sign.

a. Na^+ indicates a sodium ion with a charge of _____.

b. Cl^- indicates a chloride ion with a charge of _____.

c. Mg^{2+} indicates a magnesium ion with a charge of _____.

a. Al^{3+}
b. S^{2-}

83

a. How would an aluminum ion with a charge of +3 be indicated?
_____

b. How would a sulfide ion with a charge of −2 be indicated?
_____

loses; positive; 1;
gains; negative; 1

84

$$Na + Cl \longrightarrow Na^+ + Cl^-$$
Reaction of sodium and chlorine

When a sodium atom reacts with a chlorine atom, the sodium atom _____ (gains/loses) 1 electron to form a sodium ion with a _____ charge of _____ (how many?).

At the same time, the chlorine atom _____ (gains/loses) 1 electron to form an ion with a _____ charge of _____ (how many?).

85

Look at this magnesium atom. Its symbol is _____.

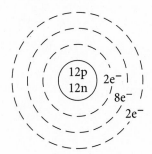

Magnesium atom

It has _____ electrons in its valence shell. So, to gain
stability, it will have to _____ (gain/lose) _____
(how many?) electrons.

The symbol for the magnesium ion is _____ (Mg^{2-}, Mg,
Mg^{2+}).

Mg; 2; lose; 2; Mg^{2+}

86

The reaction between magnesium and chlorine may be written as:

$$Mg \ + \ Cl_2 \ \longrightarrow \ Mg^{2+} \ + \ 2Cl^-$$

This reaction shows that 1 chlorine molecule reacts with 1 magnesium
atom to form 2 chloride ions, each with a charge of _____, and
1 magnesium ion with a charge of _____.

−1; +2

a. 3
b. 6
c. −2
d. S^{2-}

87

a. The sulfur atom, atomic number 16, has how many energy levels? _____

b. How many electrons are in its valence shell? _____

c. If a sulfur atom gains 2 electrons to fill its valence shell, it will have a charge of _____.

d. The symbol for the sulfide ion is _____.

88

How can we tell whether an atom will lose or gain electrons to reach a stable structure of 8 electrons in its valence shell? Look over the following diagrams.

Form positive ions

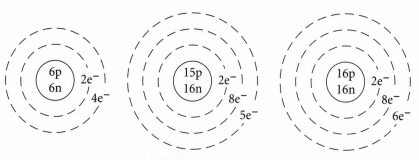

Do not form positive ions

In general, an atom with _____ (how many?) electrons in its valence shell tends to lose electrons and form a _____ (positively/negatively) charged ion.

1, 2, or 3 (or equivalent answer); positively

89

Sodium has atomic number 11. How many electrons does it have in its valence shell? _____

1; loss; Na$^+$

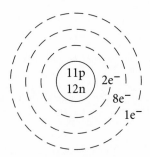

Sodium atom

The sodium ion, therefore, is formed by the _____ (loss/gain) of 1 electron.

The symbol for the sodium ion is _____.

90

+1; H$^+$

Here is the hydrogen atom, atomic number 1 and mass number 1.

Hydrogen atom

Hydrogen usually loses its 1 electron to form a hydrogen ion with a charge of _____.

The hydrogen ion can be represented as _____.

3; loss; Al^{3+}

91

Here is an aluminum atom:

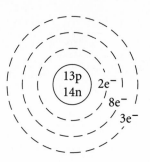

Aluminum atom

Its valence shell contains _____ electrons. The aluminum ion is formed by the _____ (loss/gain) of 3 electrons.

The symbol for the aluminum ion is _____.

Mg^{2+}

92

Magnesium, atomic number 12, forms a magnesium ion that may be represented as _____.

93

Look at the following diagrams carefully.

Form negative ions

Do not form negative ions

In general, atoms that have _____ (how many?) electrons in their valence shells form negatively charged ions.

94

When an atom gains electrons to fill its valence shell, it forms _____ charged ions.

9; 7; gain; 1; F⁻

95

Complete the following for the fluorine (F) atom shown here:

Fluorine atom

Atomic number: _____
Number of electrons in valence shell: _____
Ion formed by _____ (gain/loss) of _____ electron(s)
Symbol for ion: _____

S; 16; 32; S²⁻

96

Complete the following for the sulfur atom shown here:

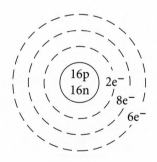

Sulfur atom

Symbol for sulfur: _____
Atomic number: _____
Mass number: _____
Symbol for ion: _____

97

Atoms with 4 or 5 electrons in their valence shells usually do not form ions. These we shall discuss later.

Atoms with 8 electrons in their valence shells are stable. Because they do not have to gain or lose electrons, you would expect these stable atoms to be:

__ reactive   __ inert (unreactive)

inert (unreactive)

98

The sodium atom, atomic number 11, has how many electrons in its valence shell? _____

The sodium atom will tend to lose the 1 electron in its valence shell to form an ion with a charge of _____.

1; +1

99

The chlorine atom, which has the symbol _____, has atomic number 17. It has how many electrons in its valence shell? _____

The chlorine atom will tend to gain _____ (how many?) electron(s) to form an ion with a charge of _____.

Cl; 7; 1; −1

IONIC BONDS

gains

100

Consider the reaction between sodium and chlorine:

$$Na + Cl \longrightarrow Na^+ + Cl^-$$

The sodium ion and the chloride ion are oppositely charged. These ions are held together by the attraction of their opposite charges. We say that there is an *ionic bond* between the sodium ion and the chloride ion.

An ionic bond is produced whenever one atom loses an electron or electrons and another atom _____ an electron or electrons.

ionic bond

101

In the reaction $Zn + S \longrightarrow Zn^{2+} + S^{2-}$, the zinc ion and the sulfur ion are held together by an _____.

ion; negatively; ion

102

When a sodium atom combines with a chlorine atom according to the equation $Na + Cl \longrightarrow Na^+ + Cl^-$, a compound containing a positively charged sodium _____ and a _____ charged chloride _____ is formed. This compound (sodium chloride) is usually written as NaCl, with the charges being understood and not written.

chloride; Cl⁻

103

When the compound NaCl is placed in water, the ionic bond holding the sodium ion and the chloride ion together is weakened, so that these ions are free to move throughout the solution.

Therefore, NaCl in water produces a *solution* containing sodium ions (Na^+) and _____ ions (_____).

ACIDS, BASES, AND SALTS

104

a. When the compound HCl is placed in water, it produces a solution containing _____ ions (_____) and _____ ions (_____).

b. When the compound H_2SO_4 is placed in water, it produces a solution containing _____ ions and HSO_4^- ions.

105

Any substance that yields hydrogen ions (H^+) in solution is called an *acid*.

$$HCl \longrightarrow H^+ + Cl^-$$
$$H_2SO_4 \longrightarrow H^+ + HSO_4^-$$
$$HNO_3 \longrightarrow H^+ + NO_3^-$$

HNO_3 is an *acid* because it yields _____ in solution.

106

A hydrogen ion is also a *proton*. Therefore, acids yield hydrogen ions or _____ in solution.

---

a. hydrogen; H^+; chloride; Cl^-
b. hydrogen (H^+)

hydrogen ions or H^+

protons

acids

$$\boxed{107}$$

Note that hydrogen ions associate with water molecules to form a "hydronium" ion $(H_3O)^+$. However, we will use the term *hydrogen ion* (H^+) in this book.

Substances that yield hydrogen ions or protons in solution are called _____. Substances that accept (react with) hydrogen ions or protons are called *bases*.

bases

$$\boxed{108}$$

When sodium hydroxide, NaOH, is placed in water, sodium ions and hydroxide ions are present:

$$NaOH \longrightarrow Na^+ + OH^-$$

The hydroxide ions (OH^-) react with hydrogen ions to form water:

$$OH^- + H^+ \longrightarrow H_2O$$

Because hydroxide ions accept (react with) hydrogen ions, OH^- ions are _____.

yes

$$\boxed{109}$$

Potassium hydroxide, KOH, also yields OH^- ions in water. Would KOH be a base? _____

bases; they accept (react with) hydrogen ions

$$\boxed{110}$$

Bicarbonate ions, HCO_3^-, are important ions in body fluids. They react as follows:

$$HCO_3^- + H^+ \longrightarrow H_2CO_3$$

Are bicarbonate ions acids or bases? _____

Why? _____

111

Ammonia is a waste product of the body's metabolism of protein. Ammonia reacts as follows:

$$NH_3 + H^+ \longrightarrow NH_4^+$$

Ammonia   Ammonium ion

Is ammonia an acid or a base? _____

base

112

Acids _____ hydrogen ions or protons.

Bases _____ hydrogen ions or protons.

yield; accept (react with)

113

When the compound NaCl is placed in water, it yields _____ ions and _____ ions.

sodium (or Na^+); chloride (or Cl^-)

114

The solution of NaCl in water would not be considered an acid because:

__ it yields OH^- ions
__ it doesn't break up into ions
__ it yields no hydrogen ions

it yields no hydrogen ions

a. salt
b. acid
c. salt
d. acid
e. base
f. acid
g. base

115

A compound that yields ions other than hydrogen ions (H^+) or hydroxide ions (OH^-) is called a *salt*. Are the following underlined substances acids, bases, or salts?

a. $\underline{Na_2CO_3} \longrightarrow 2Na^+ + CO_3^{2-}$ _____

b. $\underline{H_2SO_4} \longrightarrow H^+ + HSO_4^-$ _____

c. $\underline{MgCl_2} \longrightarrow Mg^{2+} + 2Cl^-$ _____

d. $\underline{H_2PO_4^-} \longrightarrow H^+ + HPO_4^{2-}$ _____

e. $\underline{NaOH} + H^+ \longrightarrow Na^+ + H_2O$ _____

f. $\underline{H_3PO_4} \longrightarrow H^+ + H_2PO_4^-$ _____

g. $\underline{Ca(OH)_2} + 2H^+ \longrightarrow Ca^{2+} + 2H_2O$ _____

ELECTROLYTES AND NONELECTROLYTES

116

Acids, bases, and salts are called *electrolytes*. Solutions of electrolytes conduct electricity because of the presence of ions:

$$HCl \longrightarrow H^+ + Cl^-$$
Hydrochloric
acid

Would a solution of hydrochloric acid be an electrolyte? _____

yes

117

Here's what happens when magnesium chloride is put in solution:

$$MgCl_2 \longrightarrow Mg^{2+} + 2Cl^-$$

a. Does the solution of $MgCl_2$ contain ions? _____

b. Is $MgCl_2$ an acid, a base, or a salt? _____

c. Is a solution of $MgCl_2$ an electrolyte? _____

a. yes
b. salt
c. yes

118

When sucrose, $C_{12}H_{22}O_{11}$, is placed in water, no ions are produced.

a. Would a solution of sucrose conduct electricity? _____

b. Would a solution of sucrose be an electrolyte or a nonelectrolyte?

_____

a. no
b. nonelectrolyte

119

Which of the following solutions would be electrolytes?_____

a. $MgSO_4 \longrightarrow Mg^{2+} + SO_4^{2-}$

b. alcohol $\longrightarrow$ no ions

c. $KOH \longrightarrow K^+ + OH^-$

d. $C_6H_{12}O_6$ (glucose) $\longrightarrow$ no ions

e. $HNO_3 \longrightarrow H^+ + NO_3^-$

a, c, e

ANIONS AND CATIONS

120

Consider the following electrolytic system, a battery whose two electrodes are immersed in a solution containing the indicated ions:

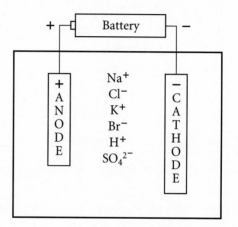

a.   The anode will attract ions of what charge? _____

b.   The cathode will attract ions of what charge? _____

a. −
b. +

121

a.   Ions attracted toward an anode are called *anions*. Anions have a _____ charge.

b.   Ions attracted toward a cathode are called *cations*. Cations have a _____ charge.

a. −
b. +

122

Which are the cations in the solution indicated in problem 120?

_____

Na^+, K^+, and H^+

123

Which are the anions in that same solution? _____

Cl^-, Br^-, and SO_4^{2-}

124

a. The principal cation inside animal cells (the principal *intra*cellular cation), the potassium ion, has a _____ (positive/negative) charge.

b. The principal cation outside animal cells (the principal *extra*cellular cation), the sodium ion, has a _____ (positive/negative) charge.

c. The iodide ion, I^-, is necessary for the proper functioning of the thyroid gland. The iodide ion is a(n) _____ (anion/cation).

a. positive
b. positive
c. anion

pH

Mathematically, pH $= -\log H^+$. A log, or logarithm, is an exponent. The log (exponent) of 10^{-3} is -3.

125

The log of 10^{-7} is _____.

-7

126

If $H^+ = 10^{-4}$, then pH $= -\log H^+ = -(-4) =$ _____.

4

a. 12
b. 1

127

That is, pH is the negative log of the H^+.

a.   If $H^+ = 10^{-12}$, pH = _____.

b.   If $H^+ = 10^{-1}$, pH = _____.

a. acid
b. neutral
c. acid

128

The acid or basic strength of a solution may be expressed in terms of a number called the *pH* of that solution. The pH scale expresses the concentration of hydrogen ions (and hydroxide ions) in solution.

The pH range is from 0 to 14, with a pH of 7 indicating a neutral solution. A pH below 7 indicates an *acid* solution.

a.   A pH of 3 indicates a(n) _____ solution.

b.   A pH of 7 indicates a(n) _____ solution.

c.   A pH of 6 indicates a(n) _____ solution.

a. 7
b. neutral

129

Water is very slightly ionized: $H_2O \longrightarrow H^+ + OH^-$. The hydrogen ion concentration (H^+) in water is 10^{-7}.

a.   What is the pH of water? _____

b.   Is water acid, basic, or neutral? _____

Although all pH values below 7 indicate acid solutions, there is a definite progression of acid strengths according to pH values.

A pH between 5 and 7 indicates a weak acid solution, between 2 and 5 indicates a moderately strong acid solution, and between 0 and 2 indicates a strong acid solution, as shown in the following chart:

130

a. Which of the following pH values indicates a solution containing a strong acid?

__ 4   __ 7   __ 1   __ 6

b. Which of the following pH values indicates a solution containing a moderately strong acid?

__ 4   __ 7   __ 1   __ 6

c. A pH of 0 indicates what strength acid solution? _____

d. A pH of 3 indicates what strength acid solution? _____

a. 1
b. 4
c. strong
d. moderately strong

131

pH values may be indicated as decimal values, as well as whole numbers. Thus, a pH of 2.56 indicates a solution whose pH lies between 2 and 3. Thus, this is a:

__ strong acid solution
__ weak acid solution
__ moderately strong acid solution

moderately strong
acid solution

6.27

132

Among the following pH values, which solution contains a weak acid?

___ 1.72   ___ 3.75   ___ 2.00   ___ 6.27

1.72

133

Among the following pH values, which solution contains a strong acid?

___ 1.72   ___ 3.75   ___ 7.00   ___ 5.00   ___ 6.38

7.00

134

Of solutions with the following pH values, which one is neutral?

___ 2.70   ___ 4.65   ___ 5.00   ___ 7.00

A solution whose pH is above 7 is called a *basic* solution.

A solution whose pH is between 7 and 9 is called a weak basic solution, between 9 and 12 a moderately strong basic solution, and between 12 and 14 a strong basic solution, as shown in the following chart:

135

Of the following pH values, which indicates a strong basic solution?

__ 3.5   __ 7.0   __ 9.5   __ 13.9   __ 6.4

13.9

136

Of the following pH values, which indicates a weak basic solution?

__ 2.7   __ 8.3   __ 4.2   __ 10.8   __ 14.0

8.3

137

Which of the following pH values indicates a moderately strong basic solution?

__ 2.2   __ 4.7   __ 11.1   __ 13.7

11.1

138

The pH of saliva lies between 5.5 and 6.9, indicating a _____ solution.

weak acid

139

Bile has a pH range of 7.8 to 8.6, so it is a _____ solution.

weak basic

weak acid

140

Urine has a pH range of 5.5 to 6.9, so it is a _____ solution.

strong acid

141

The gastric juices have a pH range of 1.6 to 1.8, so they make up a _____ solution.

weak basic

142

Blood has a pH range of 7.35 to 7.45, so it is a _____ solution.

weak basic

143

The pancreatic juices have a pH range of 7.5 to 8.0, so they make up a _____ solution.

a. 10
b. weak

144

A difference of 1 in pH is equivalent to a 10-fold difference in acid or base strength. That is, an acid of pH 2 is 10 times as strong as an acid of pH 3. Likewise, a base of pH 10 is 10 times as strong as a base of pH 9.

a. A solution of pH 4.73 is _____ times as strong as one of pH 5.73.

b. A solution of pH 9.81 is 10 times as _____ (strong/weak) as one of pH 10.81.

145

Thus a small difference in pH can represent a _____ (considerable/negligible) difference in acid or base strength.

Electrode ——

Solution ——

considerable

BUFFERS

146

Recall that acids yield hydrogen ions (H^+) or protons in solution. Both hydrochloric acid (HCl) and acetic acid (CH_3COOH), are acids, as indicated by the following reactions:

$$HCl \longrightarrow H^+ + Cl^-$$
$$CH_3COOH \rightleftharpoons CH_3COO^- + H^+$$

However, the HCl is 100% ionized, as shown by the one-way arrow, whereas CH_3COOH is less than 1% ionized, as shown by the double (equilibrium) arrows. That is, relatively speaking, HCl yields many H^+ ions in solution, whereas CH_3COOH yields few H^+ ions.

a. few
b. strong
c. weak

Hydrochloric acid, HCl, is called a *strong acid*. It yields relatively many H^+ ions in solution.

a. Acetic acid, CH_3COOH, is called a *weak acid*. It yields relatively _____ (many/few) H^+ ions in solution.

b. Sulfuric acid, H_2SO_4, yields relatively many H^+ ions in solution. It is a _____ (strong/weak) acid.

c. Citric acid, which yields relatively few H^+ ions in solution, is a _____ (strong/weak) acid.

a. many
b. few

147

A base accepts or reacts with protons (hydrogen ions).

a. A strong base reacts with relatively _____ (many/few) H^+ ions.

b. A weak base reacts with relatively _____ (many/few) H^+ ions.

a. strong
b. weak

148

a. NaOH, sodium hydroxide, reacts with relatively many H^+ ions. It is a _____ (strong/weak) base.

b. Sodium bicarbonate, $NaHCO_3$, reacts with relatively few H^+ ions. It is a _____ (strong/weak) base.

149

a.   When a strong acid is added to water, the pH of the solution should
     go _____ (up/down).

b.   When a strong base is added to water, the pH of the solution should
     go _____ (up/down).

The addition of a weak acid or a weak base to water should have very
little effect on the pH.

150

A *buffer solution* is one that maintains a constant pH upon the addition
of small amounts of either acid or base. A buffer consists of a weak acid
and a weak base. A buffer solution can "sponge up" excess H^+ if added
to a solution, or it can release H^+ if the H^+ concentration drops. One
such buffer in the body is the H_2CO_3/$NaHCO_3$ system.

Suppose that a strong acid such as HCl is added to the buffer. The reaction is:

$$HCl \: + \: NaHCO_3 \: \longrightarrow \: H_2CO_3 \: + \: NaCl$$
$$\text{Strong acid} \quad \text{Weak base} \qquad\qquad \text{Weak acid} \qquad \text{Salt}$$

Thus the strong acid has been changed to a weak acid. Should the pH
change? _____

151

If a strong base such as sodium hydroxide, NaOH, is added to the buffer,
the reaction is:

$$NaOH \: + \: H_2CO_3 \: \longrightarrow \: NaHCO_3 \: + \: H_2O$$
$$\text{Strong base} \quad \text{Weak acid} \qquad\qquad \text{Weak base} \qquad \text{Water}$$

The strong base has been changed to a _____ base. Should the
pH change? _____

it should not change it

152

Another buffer system in the body is the phosphate buffer system, which consists of NaH_2PO_4/Na_2HPO_4. What effect will the addition of a strong acid have on the pH of this buffer? _____

no

153

Should the addition of base affect the pH of a phosphate buffer system? _____

no effect

154

What effect should the removal of acid (which is equivalent in effect to the addition of base) have on the pH of a buffer solution? _____

none

155

The pH of the blood (a buffer solution) lies in the range of 7.35 to 7.45.

Normal metabolic processes add acid to the bloodstream. What effect should these acids have on the pH of the blood? _____.

yes

156

Protein buffers are also present in the body. Should they behave similarly to the bicarbonate and phosphate buffers? _____

Liquid Mixtures 5

SOLUTIONS

A solution consists of a *solute* (a relatively less abundant substance) dissolved in a *solvent* (a relatively more abundant substance). In living organisms the solvent is water.

157

a. When oxygen dissolves in water, oxygen is the _____ (solute/solvent).

b. When alcohol is dissolved in water, alcohol is the _____ (solute/solvent).

c. When salt is dissolved in water, water is the _____ (solute/solvent).

158

A solution consists of a _____ and a _____.

a. solute
b. solute
c. solvent

solute; solvent

solution

Solute + solvent ⟶ _____.

a. yes
b. homogeneous
c. transparent

Solutions are transparent and homogeneous. *Homogeneous* means "the same throughout," as compared with *heterogeneous*, which means "not the same throughout."

a. In a saltwater solution, is one part the same as any other part? _____

b. Solutions are _____ (homogeneous/heterogeneous).

c. Solutions are _____ (transparent/opaque).

a. yes
b. yes

In general, solutions pass through membranes.

a. If glucose is administered intravenously, will the glucose solution pass through the membranes in the body? _____

b. Will saline (salt) solution pass through a membrane? _____

water

For medical use, solutions are frequently described in terms of *percent* or *parts per hundred*.

In a percent solution, the percentage number indicates the number of grams of solute present in 100 mL of solution. The solvent is usually _____.

163

a.   A 5% glucose solution contains 5 g of glucose in _____ mL of solution.

b.   A 2% boric acid solution contains _____ g of boric acid in _____ mL of solution.

164

Physiologic saline solution, 0.95% NaCl solution, contains _____ g of NaCl in _____.

165

Another method of expressing solution concentration is in terms of *molarity*, or *moles* of solute per liter of solution. A *mole* of a compound is the number of grams of that compound equal to its molecular mass.

The atomic mass of hydrogen is 1 and that of oxygen, 16. The molecular mass of water, H_2O, which consists of 2H and 1O, is $(2 \times 1) + (1 \times 16)$, or _____. Therefore, one mole of water has a mass of _____ g.

166

What is the molecular mass of glucose, $C_6H_{12}O_6$? _____
Atomic masses are: C = 12, H = 1, O = 16.

a. 1; 1 liter
b. 120; 1 liter
c. 0.1; solution
d. 5.85

167

A 1-molar (1M) glucose solution contains 1 mole (180 g) of glucose dissolved in enough water to make 1 liter of solution.

Use the following atomic masses in making calculations:

H	1	O	16
Mg	24	S	32
Na	23	Cl	35.5

a. To prepare a 1M $MgSO_4$ solution, take _____ mole(s) of $MgSO_4$ and dissolve in enough water to prepare _____ (how much?) of solution.

b. To prepare a 1M $MgSO_4$ solution, take _____ g of $MgSO_4$ and dissolve in enough water to make _____ (how much?) of solution.

c. To prepare a 0.1M NaCl solution, take _____ mole(s) of NaCl and dissolve in enough water to make 1 liter of _____.

d. To prepare a 0.1M NaCl solution, take _____ g of NaCl and dissolve in enough water to make 1 liter of solution.

168

Osmolarity is a method of expressing solution concentration based on the number of particles in solution.

$$\text{Osmolarity} = \text{molarity} \times \frac{\text{number of particles}}{\text{molecule}}$$

Osmolarity is expressed in the units osmoles, or *osmol*.

Consider the following reactions:

$$NaCl \longrightarrow Na^+ + Cl^-$$

1 molecule $\longrightarrow$ 2 particles

$$Glucose \longrightarrow glucose$$

1 molecule $\longrightarrow$ 1 particle

$$K_2SO_4 \longrightarrow 2K^+ + SO_4^{2-}$$

1 molecule $\longrightarrow$ 3 particles

a. A 1-molar (1M) glucose solution has a concentration of _____ osmol.

b. A 0.5M NaCl solution has a concentration of _____ osmol.

c. A 0.4M K_2SO_4 solution has a concentration of _____ osmol.

a. 1
b. 1
c. 1.2

169

Osmolarity may also be expressed in the units milliosmoles, or *mosmol*, where

$$1 \text{ osmol} = 1000 \text{ mosmol}$$

To change osmoles to milliosmoles, multiply by 1000.

a. A 0.15M NaCl solution has a concentration of _____ osmol.

b. A 0.15M NaCl solution has a concentration of _____ mosmol.

Note that human body fluids have an osmolarity of approximately 300 milliosmol.

a. 0.30
b. 300

SUSPENSIONS

170

When sand is placed in water and shaken, a sand-water suspension is produced.

a.   When the shaking is stopped, does the sand settle? _____

b.   Does sand dissolve in water? _____

c.   In a suspension, is there a solute? _____ Why or why not?

_____

Note that particles in a suspension are so large that they settle under the influence of gravity.

a. yes
b. no
c. no; the substance does not dissolve in water

171

a.   A suspension is _____ (homogeneous/heterogeneous).

b.   A solution is _____ (homogeneous/heterogeneous).

c.   Is a suspension transparent? _____

d.   Is a solution transparent? _____

a. heterogeneous
b. homogeneous
c. no
d. yes

172

Blood is a suspension of blood cells in blood plasma. If blood is treated so that it doesn't clot and is then placed in a test tube, will the blood cells settle? _____

What causes the blood cells to settle? _____

yes; gravity

COLLOIDS
(COLLOIDAL DISPERSIONS)

Colloids consist of very tiny particles suspended in a liquid, usually water. (Colloids may also be suspended in solids and in gases.) Colloids differ from suspensions in that colloids do not settle. Colloidal dispersions are usually translucent or milky white. They do not pass through membranes.

173

One example of a colloid is protein.

a. A colloidal dispersion of protein will be _____ (transparent/translucent).

b. A colloidal dispersion of protein _____ (will/will not) settle.

c. A colloidal dispersion of protein _____ (will/will not) pass through a membrane.

a. translucent
b. will not
c. will not

174

Many proteins are colloids. Should such proteins be able to pass through the membranes of the kidneys and be present in the urine? _____

Should salt, NaCl? _____

no; yes

Diffusion and Osmosis 6

DIFFUSION

If a bottle of perfume is opened in a room, the odor is soon apparent throughout the room. The molecules of perfume diffuse, or spread out, into the available volume of air.

Next, suppose a sugar solution is placed on one side of a permeable (nonselective) membrane and pure water is placed on the other side of that membrane.

o = sugar molecules

The sugar molecules (the solute) will diffuse across the permeable membrane until the concentrations on both sides of the membrane are equal.

When this occurs, sugar molecules will continue to diffuse across the membrane in both directions so that there will be no net change in concentration.

175

During diffusion, solute particles move from:

__ high concentration of solute to lower concentration of solute
__ low concentration of solute to higher concentration of solute

high concentration of solute to lower concentration of solute

176

In passive transport, solute particles flow from a _____ (high/low) concentration to one of _____ (higher/lower) concentration.

high; lower

177

Can diffusion occur across a permeable membrane? _____

yes

178

Can diffusion occur without a membrane? _____

yes

179

Diffusion across a biologic membrane is called _____.

passive transport

180

Diffusion is the flow of _____ (solute/solvent) molecules.

solute

OSMOSIS

181

Do all solutions have the same solute concentration? _____

no

A solution that has a higher solute concentration than another solution is said to be *hypertonic* (hyperosmotic) when compared with the second solution.

A solution that has a lower solute concentration than another solution is said to be *hypotonic* (hypoosmotic) when compared with the second solution.

A solution that has the same solute concentration as another solution is said to be *isotonic* (isoosmotic) when compared with the second solution.

Look at the following illustrations of solutions whose solute is indicated by small circles.

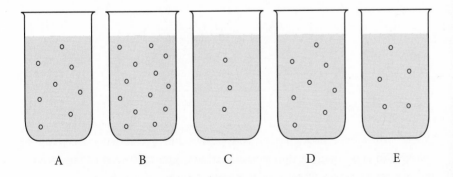

    A   B   C   D   E

a. C, E
b. B
c. D

nonselective

182

a.   Solution A is hypertonic to solution(s) _____ .

b.   Solution A is hypotonic to solution(s) _____ .

c.   Solution A is isotonic to solution(s) _____ .

183

A permeable membrane is _____ (selective/nonselective).

A semipermeable membrane is one that is selectively permeable.

Consider the following system in which a selectively permeable membrane separates a dilute (weak) sugar solution from a concentrated (strong) sugar solution. We will assume that this semipermeable membrane is permeable to water but not to sugar molecules.

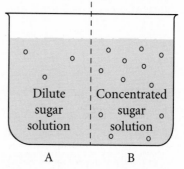

Water will flow through the semipermeable (selectively permeable) membrane and equalize the concentrations. Such a flow is called *osmosis* and is a special case of passive transport. The results will be:

Selectively permeable membrane

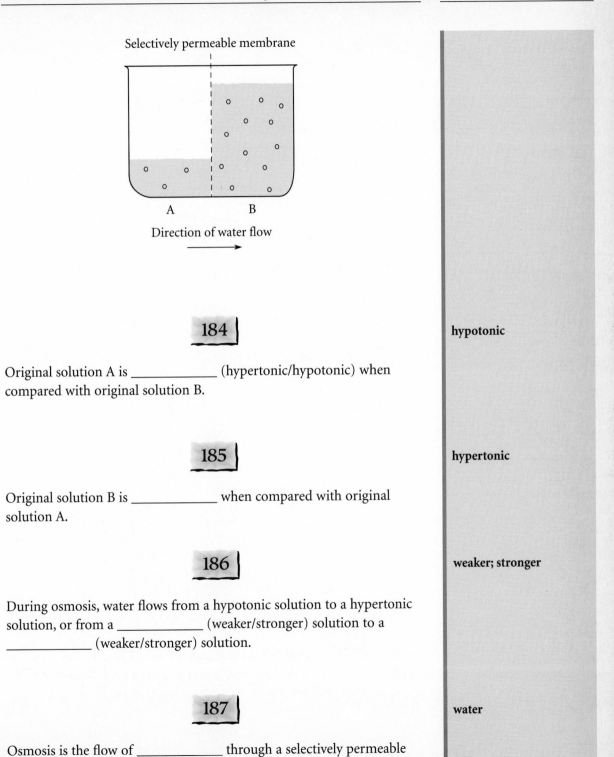

A   B

Direction of water flow

⎡ 184 ⎤

Original solution A is _____ (hypertonic/hypotonic) when compared with original solution B.

⎡ 185 ⎤

Original solution B is _____ when compared with original solution A.

⎡ 186 ⎤

During osmosis, water flows from a hypotonic solution to a hypertonic solution, or from a _____ (weaker/stronger) solution to a _____ (weaker/stronger) solution.

⎡ 187 ⎤

Osmosis is the flow of _____ through a selectively permeable membrane.

hypotonic

hypertonic

weaker; stronger

water

yes

188

Will osmosis continue until the two solutions have equal concentrations? _____

solute; water molecules

189

Diffusion is the flow of _____ particles; osmosis is the flow of _____ through a selectively permeable (semipermeable) membrane.

hypertonic (concentrated); hypotonic (dilute)

190

Diffusion is the flow of solute particles from a _____ solution to a _____ solution.

hypotonic (dilute); hypertonic (concentrated)

191

Osmosis is the flow of water through a membrane from a _____ solution to a _____ solution.

yes; yes

192

Is diffusion passive transport? _____ Is osmosis? _____

193

Consider the following system where dilute and concentrated salt solutions are separated by a selectively permeable membrane that allows the salt ions and water to flow through.

Selectively permeable membrane

Dilute salt solution

Concentrated salt solution

A

B

a. What term is used to compare the strength of solution A with that of solution B? _____

b. What term is used to compare the strength of solution B with that of solution A? _____

c. When diffusion takes place, the flow of solute will be from _____ to _____ solution.

a. hypotonic
b. hypertonic
c. hypertonic (concentrated); hypotonic (dilute)

194

Such a flow of solute is also called _____ transport.

passive

195

While diffusion is taking place, osmosis can also occur.

During osmosis, _____ flows through a membrane from a _____ solution to a _____ solution.

water; hypotonic; hypertonic

no (opposite directions)

Selectively permeable membrane

Osmosis →

← Diffusion of ions

Dilute salt solution | Concentrated salt solution

| 196 |

Does osmosis take place in the same direction through the membrane as diffusion of a solute? _____

| 197 |

In the following diagram indicate the direction of flow of osmosis and diffusion of ions.

Selectively permeable membrane

Dilute salt solution | Concentrated salt solution

OSMOTIC PRESSURE

| 198 |

Consider the following system in which water is placed on both sides of a U-tube that has a selectively permeable membrane (permeable to water but not to solutes) dividing the two halves.

Will osmosis take place? _____ Why or why not? _____

_____

199

Next, consider a similar U-tube in which water is placed on one side and sugar solution on the other. The membrane is impermeable to sugar but permeable to water.

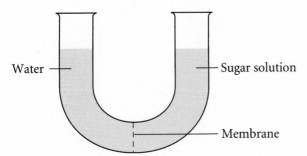

a.   Will osmosis take place? _____

b.   In which direction will osmosis take place? _____

The net result will be a rise in the liquid level on the sugar side and a drop in the liquid level on the water side, as in the following figure:

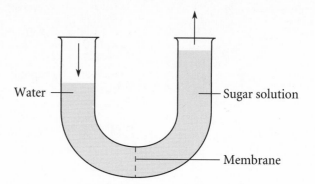

If a piston is placed on the sugar side of the U-tube and pressure is applied to that piston, then osmosis can be prevented. The amount of pressure required to counteract osmosis is called the *osmotic pressure*.

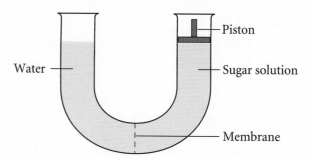

In general, the greater the difference in concentration on two sides of a membrane, the greater the osmotic pressure.

C; B

$$\underline{200}$$

In the following diagrams of sugar solutions separated from each other by a membrane impermeable to sugar but permeable to water, which will exert the greatest osmotic pressure? _____ The least? _____

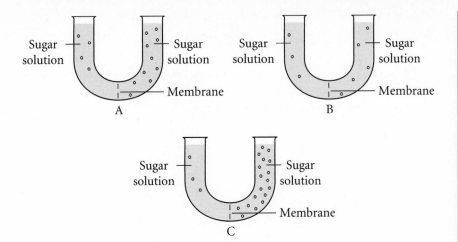

A

B

C

201

Consider the following system where a U-tube has water on one side of a membrane and a dilute NaCl solution on the other. The membrane is permeable to sodium and chloride ions and to water.

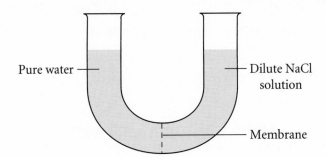

a. Will osmosis take place? _____

b. In which direction? _____

c. Will diffusion take place? _____

d. In which direction? _____

e. Will the two concentrations equalize? _____

a. yes
b. into NaCl solution
c. yes
d. into the water
e. yes

a. yes
b. yes
c. yes
d. yes
e. diffusion, or passive
   transport

202

Now consider a U-tube with NaCl solution on one side and KCl solution on the other; the solutions have the same concentrations. The membrane is equally permeable to both NaCl and KCl.

NaCl solution —| |— KCl solution

—— Membrane

a.   Will NaCl diffuse? _____

b.   Will KCl diffuse? _____

c.   Will the concentrations of KCl be the same on each side when diffusion is finished? _____

d.   Will the concentrations of NaCl be the same on each side? _____

e.   What is this motion of solute particles called? _____

ACTIVE TRANSPORT

203

Next consider an animal cell with cation concentrations in millimoles per liter as listed in the following table.

Cation	Intracellular	Extracellular
Na	10	140
K	140	4

The same situation is shown in the following diagram, where the size of the symbol indicates relative concentration.

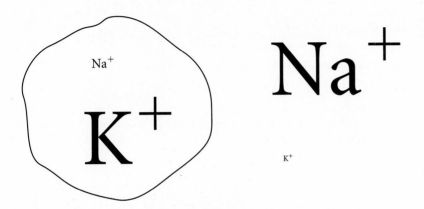

If diffusion or passive transport were the only factor affecting these ions, would the concentrations tend to equalize? _____

yes

204

Since the extracellular and intracellular concentrations of sodium and potassium ions are unequal in animal cells, there must be another force acting (shown in the following diagram).

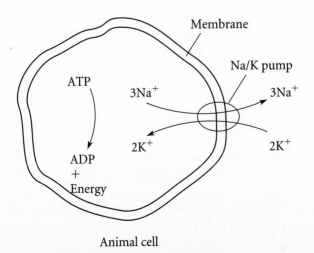

Animal cell

a. ATP
b. Na^+
c. K^+
d. no
e. 2

This type of movement is called *active transport*.

a.   What energy compound is required for active transport?

    _____

b.   What ion does active transport pump out of the cell? _____

c.   What ion does active transport pump into the cell? _____

d.   Are the movements of Na^+ and K^+ equal in amount? _____

e.   For every three sodium ions pumped out of a cell, how many potassium ions are pumped in? _____

205

active; ATP

This type of ion movement is called _____ transport and requires the energy compound _____.

206

no

This ion system pumping sodium ions out of a cell and potassium ions into a cell is called a *sodium-potassium pump*.

Are Na^+ and K^+ pumped equally across a membrane? _____

207

3, 2

How many Na^+ are pumped out and how many K^+ are pumped in?

_____

208

less

Because both Na^+ and K^+ are positively charged ions, 3 positive charges are pumped out of a cell for every 2 that are pumped in. Should the inside of the cell be more or less positive than the outside of the cell?

_____

209

Should the inside of the cell be positive or negative when compared with the outside of the cell? _____

210

Animal cells have a *membrane potential* or a small voltage difference between the inside and outside of the cell.

An animal cell has a small voltage difference or _____ between the inside and outside of the cell.

211

Complete the following diagram.

_____ transport

Active transport

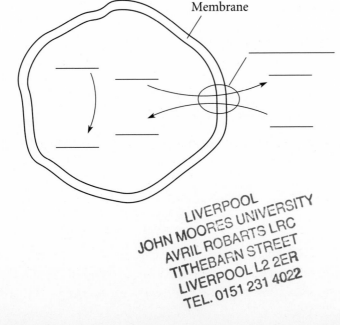

Membrane

DEPOLARIZATION AND REPOLARIZATION

212

A resting neuron has a higher _____ (K^+/Na^+) concentration inside the cell and a higher _____ (K^+/Na^+) ion concentration outside that cell.

213

A resting cell has a $(+/-)$ charge inside.

214

A resting neuron has a fairly constant electrical potential difference (voltage) of -50 to -100 millivolts inside. The following diagram represents a resting neuron. Note that the membrane potential is between -80 and -100 mV. Does a resting neuron have a positive or negative electrical potential inside? _____

K^+; Na^+

$-$

negative

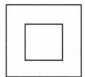

215

Some cells, particularly nerve and muscle cells, can generate changes in the electrical potential difference across their cell membranes when a stimulus is applied. Such cells have sodium and potassium channels that can open to allow these ions in or out of the cell. When closed, these channels do not allow ions to pass through.

A sodium channel may be represented by a square shape (S for sodium; S for square). The diagram on the left represents an open sodium channel, and the diagram on the right represents a closed sodium channel.

If a triangular shape represents a potassium channel, what do the following diagrams represent? a. _____, b. _____

a   b

a. **closed potassium channel**
b. **open potassium channel**

216

The following diagram illustrates a resting neuron. Which channels are open? _____ Closed? _____

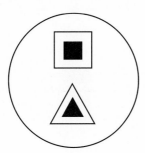

217

If both Na and K channels are closed, should the intracellular Na^+ and K^+ concentrations remain constant?

218

If a neuron is stimulated, the sodium channels open but the potassium channels remain closed, as in the following diagram.

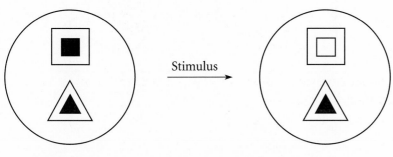

Resting cell

a. Will sodium ions flow into or out of the cell? _____

b. Will potassium ions flow into or out of the cell? _____

+

a stimulus

resting state

219

Sodium ions have what type of charge, + or −? _____

As sodium ions flow into the cell, the potential of that cell will change from − to +. The cell is said to be *depolarized*.

220

What causes the sodium channels to open? _____

221

As the cell becomes depolarized, the membrane potential changes from − to +. Section 2 of the following diagram illustrates depolarization. What does section 1 indicate? _____

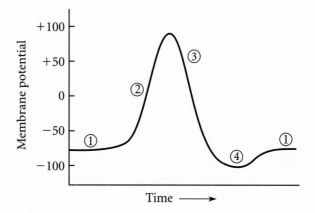

222

Color in the following diagram to show which channels should be open and which should be closed. What is effect *A*?

Resting cell   Depolarization

Resting cell

↓*A*

Depolarization

A represents a stimulus

neither; out

223

Once the cell has become depolarized, the sodium channels close and the potassium channels open.

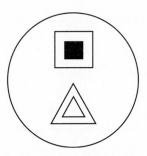

Will sodium ions flow into or out of the cell? _____

Will potassium ions flow into or out of the cell? _____

it becomes negative

This phase is called *repolarization*, as shown in section 3 of the following diagram.

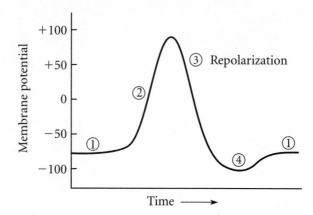

During repolarization, as potassium ions flow out of the cell, what happens to the membrane potential? _____

resting state; depolarization

What is section 1 called in the problem 224 diagram? _____

What is section 2 called? _____

Color the following diagram to indicate which of the following channels are open or closed.

Resting cell   Depolarization

Repolarization

Resting cell

Stimulus

Depolarization

Repolarization

The flow of potassium ions out of the cell during repolarization con-tinues afterward because these channels are slow to close. This produces an *undershoot* or greater negative charge inside the cell than was present in the resting state, as shown in the following diagram.

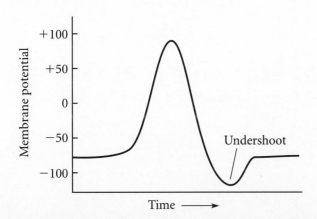

Finally, both channels are closed and active transport takes over to return the cell to its resting state.

a. sodium
b. potassium
c. 3 Na^+ out; 2 K^+ in
d. potassium
e. sodium

227

a. During active transport, _____ ions are carried out of the cell.

b. During active transport, _____ ions are carried into the cell.

c. What are the relative number of Na^+ and K^+ ions moved out of and into the cell? _____

d. During undershoot, which channels are slow to close?
    _____

d. During undershoot, which channels remain closed? _____

228

1: resting state
2: depolarization
3: repolarization
4: undershoot
5: resting state

In the following diagram, label sections 1 through 5.

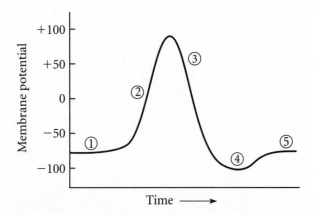

229

In the following diagram, indicate the charges inside the cell membrane. Also mark which channels are open and which are closed. Finally, label *A* and *B*.

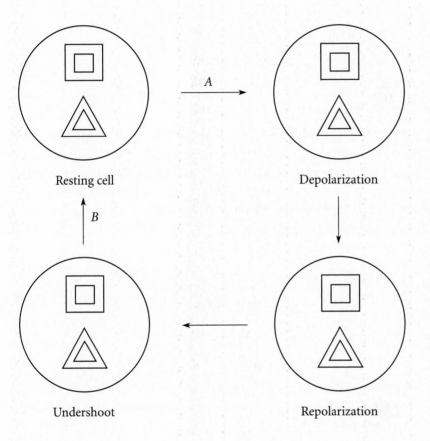

Resting cell

A

Depolarization

B

Undershoot

Repolarization

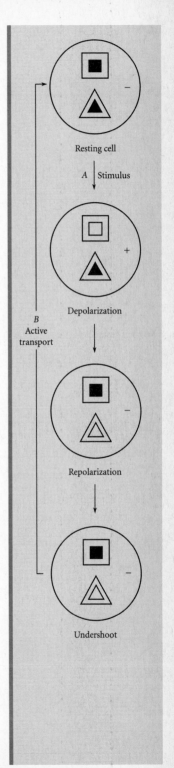

Resting cell

A | Stimulus

Depolarization

B
Active transport

Repolarization

Undershoot

PROPAGATION OF A NERVE IMPULSE

Nerve impulses are propagated in an axon by means of a series of depolarizations and repolarizations, such as indicated in the following drawing:

Resting cell   Repolarization   Depolarization

How does this occur? Depolarization increases the membrane potential to a critical level called the threshold potential. This in turn triggers an *action potential*, or nerve impulse, which is transmitted to an adjacent cell. Consider the following diagram of the cross section of a small segment of an axon that illustrates three adjacent cells in their resting state with a negative potential inside each cell.

When a stimulus is applied to cell A, it causes that cell to become depolarized or positively charged in its inner membrane. This produces an action potential.

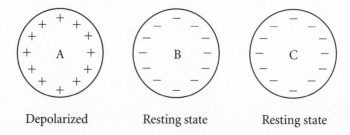

Depolarized   Resting state   Resting state

The depolarization of cell A in turn causes the depolarization of cell B so that the action potential (nerve impulse) travels along the indicated path.

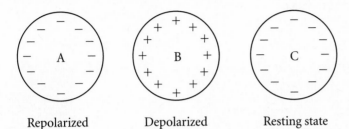

Repolarized   Depolarized   Resting state

230

While the impulse is traveling from A to B, cell A becomes repolarized and returns to its resting state.

The depolarization of cell B will have what effect on cell C? _____

_____

it will become depolarized

231

What happens to the action potential (nerve impulse)? _____

_____

it travels from B to C

232

What happens to cell B after the impulse is transmitted to C? _____

_____

it becomes repolarized and then returns to its resting state

This effect may be considered as similar to that of a section of standing dominoes. When the first one is pushed over, the motion is transmitted to succeeding ones just as a nerve impulse travels from one end of an axon to the other.

Axon   Axon

However, note that the last falling domino is too far from the next one to cause it to fall. The effect then stops.

Similarly, the movement of a nerve impulse travels along an axon to the end but cannot jump across the gap to the next axon. Something must carry that impulse across the gap.

no

233

The following diagram illustrates the end of one axon and the beginning of the succeeding one.

Can the nerve impulse jump across the synaptic cleft? _____

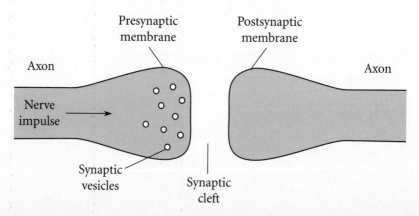

Presynaptic membrane   Postsynaptic membrane

Axon   Axon

Nerve impulse →

Synaptic vesicles   Synaptic cleft

234

When the action potential reaches the presynaptic membrane, it depolarizes it. The depolarization of the presynaptic membrane triggers an influx of calcium ions, which in turn causes the synaptic vesicles to fuse with the presynaptic membrane.

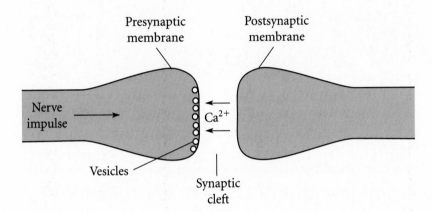

What effect does a nerve impulse have on the presynaptic membrane?

_____

235

The depolarization of the presynaptic membrane causes the influx of which ions? _____

236

The influx of calcium ions has what effect upon the vesicles?

_____

it depolarizes it

calcium

they fuse with the presynaptic membrane

When the vesicles fuse to the presynaptic membrane, they release neurotransmitters that can travel across the synaptic cleft to the postsynaptic membrane.

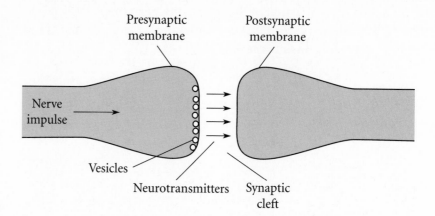

At the postsynaptic membrane, the neurotransmitters bind to receptor sites, which then open specific ion channels in the postsynaptic membrane. (Which ion channel is opened depends on the identity of the neurotransmitter.)

yes

237

When an ion channel is open, can ions move across a membrane?

_____

no

238

When an ion channel is closed, can ions move across a membrane?

_____

neurotransmitter

239

When the vesicles bind to the presynaptic membrane, what type of substance is released? _____

240

When a neurotransmitter is released, where does it go? _____

into the synaptic cleft

241

When the neurotransmitter reaches the postsynaptic membrane, what effect does it have on receptor sites? _____

causes ion channels to open

242

Once the ion channels open, selected ions (which ion depends on the identity of the neurotransmitter) enter the postsynaptic membrane, causing depolarization and thus a new action potential there. The new action potential then carries the nerve impulse along that axon.

When a neurotransmitter reaches its receptor site on the postsynaptic membrane, it opens ion channels for which ions? _____

depends on identity of the neurotransmitter

243

When the ion channels are open, selected ions can enter the postsynaptic membrane, causing _____.

depolarization

244

When the postsynaptic membrane is depolarized, what effect is produced? _____

an action potential

it begins sending a nerve impulse along that axon

245

What effect does this action potential have on the second axon?

After the postsynaptic membrane is depolarized, enzymes degrade the neurotransmitters so that the membrane is ready for succeeding signals from new molecules of neurotransmitters.

246

A: presynaptic membrane
B: synaptic cleft
C: postsynaptic membrane
D: vesicles
E: ion channels
F: action potential
G: flow of neurotransmitters

Label A through G in the following diagram.

PART **II**

Organic
Chemistry

The Covalent Bond 8

247

You've already learned that atoms tend to lose electrons to form positively charged ions if they have 1, _____, or _____ electrons in their valence shells.

On the other hand, atoms that have _____ or _____ electrons in their valence shells tend to gain electrons to form _____ charged ions.

2; 3; 6; 7; negatively

248

Atoms with 4 or 5 electrons in their valence shells tend to *share* electrons. When atoms share electrons, they form a *covalent bond*.

Which of the following illustrated atoms would form covalent bonds?

_____

C and N

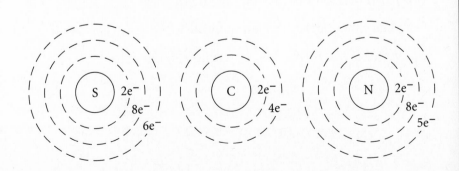

a. Mg, H, Al
b. Cl, S
c. Ne
d. stable and unreactive

249

Answer the following questions about these illustrated atoms.

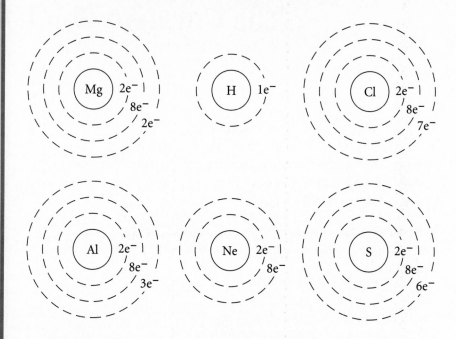

a. Which of the atoms will lose electron(s) to form positively charged ions? _____

b. Which of the atoms will gain electron(s) to form negatively charged ions? _____

c. Which of the atoms has a complete valence shell? _____

d. You know that elements with a complete valence shell are usually:

___ unstable and reactive   ___ stable and unreactive

250

a.   When two atoms share electrons, they are held together by a
_____ bond.

b.   When two atoms share electrons, no ions are produced. Is the
resulting compound an electrolyte or a nonelectrolyte? _____

a. covalent
b. nonelectrolyte

251

Atoms with 4 or 5 electrons in their valence shells tend to share electrons, but under some conditions atoms with more or fewer electrons
in their valence shells may also share electrons.

When one atom loses an electron and another atom gains that
electron, the ions thus formed are held together by a(n) _____
(covalent/ionic) bond.

ionic

252

a.   Compounds containing ionic bonds are _____
(electrolytes/nonelectrolytes).

b.   Compounds containing only covalent bonds are _____
(electrolytes/nonelectrolytes).

a. electrolytes
b. nonelectrolytes

4

253

The carbon atom, atomic number 6, has 4 electrons in its valence shell. It thus needs _____ more electrons to become stable. It can get these electrons by sharing. The 4 electrons in the valence shell of the carbon atom may be represented as here, with one dot for each electron in the valence shell.

$$\cdot \ \overset{\displaystyle \cdot}{\underset{\displaystyle \cdot}{C}} \ \cdot$$

2

254

When atoms share electrons, they always try to reach stable configurations of 8 electrons in the valence shells. The only exception to this rule is the hydrogen atom, which reaches a stable configuration when it has 2 electrons in its valence shell, which is also its only energy level. This first energy level can hold only _____ electrons.

4

255

Consider the carbon atom with 4 electrons in its valence shell. With how many hydrogen atoms can it share electrons to reach a stable configuration of 8 electrons in its valence shell? _____

256

8; 2

The four hydrogen atoms can share electrons with a carbon atom to form a compound of the following type:

There are _____ electrons around the carbon atom and _____ electrons around each hydrogen atom.

257

2

When two hydrogen atoms combine, they share electrons. Each hydrogen atom then has _____ electrons around it.

258

4

In the following compound, what is the total number of covalent bonds? _____

$$
\begin{array}{c}
\text{H} \\
\cdot\cdot \\
\text{H} : \text{C} : \text{H} \\
\cdot\cdot \\
\text{H}
\end{array}
$$

7

259

The chlorine atom, atomic number 17, has _____ electrons in its valence shell.

The electrons in the valence shell of the chlorine atom may be represented as

$$\cdot \; \overset{\displaystyle ..}{\underset{\displaystyle ..}{Cl}} \; :$$

When two chlorine atoms combine, they share electrons to reach a stable configuration of 8 electrons in the valence shell of each atom, or

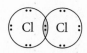

covalent

260

What type of bond is there between two chlorine atoms? _____

261

A covalent bond is frequently indicated by a short line rather than by dots; thus, the hydrogen molecule may be represented as H : H or H — H. The short line indicates a pair of shared electrons.

The compound CH_4 may be represented as

$$\overset{\displaystyle H}{\underset{\displaystyle H}{H : \overset{..}{\underset{..}{C}} : H}} \quad \text{or} \quad \overset{\displaystyle H}{\underset{\displaystyle H}{H - \overset{|}{\underset{|}{C}} - H}}$$

Give the second representation of the chlorine molecule, Cl_2.

Cl— Cl

$$\overset{\displaystyle ..}{\underset{\displaystyle ..}{:\ Cl}}\ \ \overset{\displaystyle ..}{\underset{\displaystyle ..}{Cl\ :}}\ \ or$$

262

a. The carbon atom, atomic number 6, needs how many electrons to complete its valence shell? _____

b. Because the carbon atom has 4 electrons in its valence shell, and because it will tend to share these electrons, how many covalent bonds will a carbon atom form? _____

a. 4
b. 4

The carbon atom, therefore, must have four covalent bonds, or must have four bonds attached to it. These bonds may be indicated as follows:

$$-\overset{\displaystyle |}{\underset{\displaystyle |}{C}}-\qquad -\overset{\displaystyle |}{C}=\qquad -C\equiv\qquad =C=$$

Note that each carbon atom has four bonds attached to it, regardless of how the bonds are arranged.

263

1

The hydrogen atom has 1 electron in its outer energy level, the first energy level. When the hydrogen atom shares electrons, how many more does it need to complete that valence shell? _____

a. 6
b. 2
c. 2

264

a. The oxygen atom, atomic number 8, has _____ electrons in its valence shell.

b. When the oxygen atom shares electrons, how many more electrons does it need to complete its valence shell? _____

c. Therefore, each oxygen atom must have how many bonds attached to it? _____

```
    H
    |
H — C — H
    |
    H
```

265

Draw the structure of the compound formed when four hydrogen atoms form bonds with a central carbon atom.

C

a. 3
b. 3

266

The arrangement of carbon atoms C — C indicates a bond or a shared electron pair between two carbon atoms.

a. How many hydrogen atoms may be attached to the left carbon atom? _____

b. How many hydrogen atoms may be attached to the right carbon atom? _____

267

a. Diagram all the hydrogen atoms attached to this structure:

$$C - C$$

b. How many bonds does each carbon atom have? _____

c. How many bonds does each hydrogen atom have? _____

a.
$$\begin{array}{c} \quad\ \text{H}\ \ \ \text{H} \\ \quad\ |\ \ \ \ | \\ \text{H}-\text{C}-\text{C}-\text{H} \\ \quad\ |\ \ \ \ | \\ \quad\ \text{H}\ \ \ \text{H} \end{array}$$

b. 4
c. 1

268

Consider this arrangement of carbon atoms: $C - C - C$

a. How many hydrogen atoms may be attached to the left carbon atom? _____

b. How many hydrogen atoms may be attached to the center carbon atom? _____

c. How many hydrogen atoms may be attached to the right carbon atom? _____

a. 3
b. 2
c. 3

269

Diagram the compound containing three attached carbon atoms, indicating all the hydrogen atoms connected to those carbon atoms.

$$C - C - C$$

$$\begin{array}{c} \quad\ \text{H}\ \ \ \text{H}\ \ \ \text{H} \\ \quad\ |\ \ \ \ |\ \ \ \ | \\ \text{H}-\text{C}-\text{C}-\text{C}-\text{H} \\ \quad\ |\ \ \ \ |\ \ \ \ | \\ \quad\ \text{H}\ \ \ \text{H}\ \ \ \text{H} \end{array}$$

a. 3
b. 3
c. 3
d. 3
e. 0 (or none)

270

Consider this arrangement of carbon atoms:

a. How many hydrogen atoms may be attached to the right carbon atom? _____

b. How many hydrogen atoms may be attached to the left carbon atom? _____

c. How many hydrogen atoms may be attached to the upper carbon atom? _____

d. How many hydrogen atoms may be attached to the lower carbon atom? _____

e. How many hydrogen atoms may be attached to the center carbon atom? _____

271

a. Diagram the structure of the compound containing the following arrangement of carbon atoms, indicating all the hydrogen atoms.

```
        C
        |
    C — C — C
        |
        C
```

a.

```
            H
            |
        H — C — H
      H     |     H
      |     |     |
  H — C —— C —— C — H
      |     |     |
      H     |     H
        H — C — H
            |
            H
```

b. It already has four bonds.

b. Why can there be no hydrogen atoms attached to the center carbon atom? _____

272

Consider the following arrangement of carbon atoms: C═C.

There are two bonds (a double bond) between the carbon atoms. Each bond (a single bond) represents 1 pair of shared electrons. A double bond represents _____.

2 pairs of shared electrons

273

a. To become stable, a carbon atom must form how many covalent bonds? _____

b. In the structure C═C, how many hydrogen atoms may be attached to the left carbon atom? _____

c. In the space provided, diagram the structure of the compound containing two carbon atoms connected by a double bond. Show all of the hydrogen atoms.

a. 4
b. 2
c.

$$
\begin{array}{c}
\text{H}\quad\text{H}\\
|\quad\ \ |\\
\text{H}-\text{C}═\text{C}-\text{H}
\end{array}
$$

When you have completed your diagram, check it by counting the number of bonds around each carbon atom.

a. 2
b. 1
c. 3

H—C=C—C—H structure with H, H, H on top and H on bottom (answer to 275)

Cl—C—Cl structure with Cl on top and Cl on bottom (answer to 276)

274

Consider the following arrangement of carbon atoms: C=C—C

a. How many hydrogen atoms may be attached to the left carbon atom? _____

b. How many hydrogen atoms may be attached to the center carbon atom? _____

c. How many hydrogen atoms may be attached to the right carbon atom? _____

275

Complete the structure of the following compound, showing all the hydrogen atoms.

$$C=C-C$$

276

Diagram the structure of carbon tetrachloride, CCl_4, in which each chlorine atom has one bond attached to it.

277

Diagram the structure of chloroform, $CHCl_3$.

$$
\begin{array}{c}
Cl \\
| \\
Cl-C-H \\
| \\
Cl
\end{array}
$$

278

Nitrogen, atomic number 7, has 5 electrons in its valence shell and so must have three bonds attached to it. Diagram the ammonia molecule, NH_3.

$$
\begin{array}{c}
H \\
| \\
N-H \\
| \\
H
\end{array}
$$

Note that the following structure

$$
\begin{array}{c}
\ \ \overset{\displaystyle H}{|}\ \ \ \ \overset{\displaystyle H}{|}\ \ \ \ \overset{\displaystyle H}{|}\ \ \ \ \overset{\displaystyle H}{|} \\
H-C-C-C-C-H \\
\ \ \underset{\displaystyle H}{|}\ \ \ \ \underset{\displaystyle H}{|}\ \ \ \ \underset{\displaystyle H}{|}\ \ \ \ \underset{\displaystyle H}{|}
\end{array}
$$

may be simplified and written as $CH_3CH_2CH_2CH_3$. Likewise,

$$
\begin{array}{c}
\ \ \overset{H}{|}\ \ \ \ \overset{H}{|}\ \ \ \ \overset{H}{|}\ \ \ \ \overset{H}{|}\ \ \ \ \overset{H}{|} \\
H-C-C-C-C-C-H \\
\ \ \underset{H}{|}\ \ \ \ |\ \ \ \ \underset{H}{|}\ \ \ \ \underset{H}{|}\ \ \ \ \underset{H}{|} \\
\ \ \ \ \ \ \ H-C-H \\
\ \ \ \ \ \ \ \ \ \ |\ \ \\
\ \ \ \ \ \ \ \ \ \ H
\end{array}
$$

may be simplified and written as $CH_3CH(CH_3)CH_2CH_2CH_3$, and

$$
\begin{array}{c}
\ \ \overset{H}{|}\ \ \ \ \overset{H}{|}\ \ \ \ \overset{H}{|}\ \ \ \ \overset{H}{|} \\
H-C-C=C-C-H \\
\ \ \underset{H}{|}\ \ \ \ \ \ \ \ \ \ \ \ \ \underset{H}{|}
\end{array}
$$

may be simplified and written as $CH_3CH{=}CHCH_3$.

279

How may the following structures be written in simplified form?

a.

$$H-\overset{\overset{\displaystyle H}{|}}{\underset{\underset{\displaystyle H}{|}}{C}}-\overset{\overset{\displaystyle H}{|}}{\underset{\underset{\displaystyle H}{|}}{C}}-H$$

_____

b.

$$H-\overset{\overset{\displaystyle H}{|}}{\underset{\underset{\displaystyle H}{|}}{C}}-\overset{\overset{\displaystyle H}{|}}{\underset{\underset{\displaystyle H}{|}}{C}}-\overset{\overset{\displaystyle H}{|}}{\underset{\underset{\displaystyle H-C-H}{|}}{C}}-\overset{\overset{\displaystyle H}{|}}{\underset{\underset{\displaystyle H}{|}}{C}}-\overset{\overset{\displaystyle H}{|}}{\underset{\underset{\displaystyle H}{|}}{C}}-H$$

H—C—H
|
H

_____

c.

$$H-\overset{\overset{\displaystyle H}{|}}{\underset{\underset{\displaystyle H}{|}}{C}}-C\equiv C-H$$

_____

d.

H
|
H—C—H
|
$$H-\overset{\overset{\displaystyle H}{|}}{\underset{\underset{\displaystyle H}{|}}{C}}-C-\overset{\overset{\displaystyle H}{|}}{\underset{\underset{\displaystyle H}{|}}{C}}-\overset{\overset{\displaystyle H}{|}}{\underset{\underset{\displaystyle H}{|}}{C}}-H$$
|
H—C—H
|
H

_____

a. CH_3CH_3
b. $CH_3CH_2CH(CH_3)CH_2CH_3$
c. $CH_3C\equiv CH$
d. $CH_3C(CH_3)_2CH_2CH_3$

The structure for benzene, C_6H_6, may be represented in two different ways, depending on the arrangement of the electrons:

Such structures are called *resonance* structures and are indicated by a double-headed arrow.

Often, either of these resonance structures is used to represent benzene. However, for simplicity, benzene is usually abbreviated as

280

What is the abbreviated structure for the following?

a.

a.

b.

H—C and ring structure with OH, CH₂CH₃ groups (2-ethylphenol type structure)

$$
\begin{array}{c}
\text{H}\\
|\\
\text{C}\\
\end{array}
$$

(benzene ring) with substituents $-OH$ and $-CH_2CH_3$

281

Draw the complete structure for each of the following simplified formulas:

a. $CH_3CH_2CH_3$

b. $CH_3CH=CHCH_3$

c. (benzene ring)—Cl

d. $CH_3CH(CH_3)CH_2CH_2CH_3$

b. (benzene ring)—OH, —CH₂CH₃

a.
$$
\begin{array}{ccccc}
 & \text{H} & \text{H} & \text{H} & \\
 & | & | & | & \\
\text{H}— & \text{C}— & \text{C}— & \text{C} & —\text{H}\\
 & | & | & | & \\
 & \text{H} & \text{H} & \text{H} &
\end{array}
$$

b.
$$
\begin{array}{ccccc}
 & \text{H} & \text{H} & \text{H} & \text{H}\\
 & | & | & | & |\\
\text{H}— & \text{C}— & \text{C}= & \text{C}— & \text{C}—\text{H}\\
 & | & & & |\\
 & \text{H} & & & \text{H}
\end{array}
$$

c.
(benzene ring with Cl substituent — chlorobenzene)

or

(benzene ring with Cl substituent — alternate resonance drawing)

d.
$$
\begin{array}{ccccc}
\text{H} & \text{H} & \text{H} & \text{H} & \text{H}\\
| & | & | & | & |\\
\text{H}—\text{C}— & \text{C}— & \text{C}— & \text{C}— & \text{C}—\text{H}\\
| & | & | & | & |\\
\text{H} & & \text{H} & \text{H} & \text{H}
\end{array}
$$
$$
\begin{array}{c}
\text{H}—\text{C}—\text{H}\\
|\\
\text{H}
\end{array}
$$

Polar and Nonpolar Covalent Bonds 9

282

Electronegativity is the attraction of an atom for electrons. The following table indicates the electronegativities of several elements.

F - 4.0	Br 2.8	Al 1.5
O 3.5	C 2.5	Ca 1.0
Cl 3.1	H 2.1	Na 0.9
N 3.0		

Note that nonmetals have higher electronegativities than do metals. The difference in atoms' electronegativities determines the type of bond that can form between them.

When atoms of two nonmetals combine, the difference in electronegativities is small or zero and a covalent bond results.

When atoms of a metal and a nonmetal react, the difference in electronegativities is relatively larger and an ionic bond results.

a. Which of these elements has the greatest attraction for electrons?

   _____

b. Which of these elements has the least attraction for electrons?

   _____

a. **F**
b. **Na**

283

In the compound Cl — Cl or Cl:Cl, there is a pair of shared electrons between the chlorine atoms. Is there a difference in electronegativity between the two chlorine atoms? _____

no

a. Cl
b. closer to

284

Because the two chlorine atoms have the same electronegativity, the electrons should be shared equally between them. That is, the electrons should not be closer to one chlorine atom than to the other. Such a bond is called a *nonpolar covalent bond*, and such a compound is called a *nonpolar compound*. Nonpolar means there is no negative end and no positive end to the bond or to the molecule.

a. In the covalent compound HCl, H — Cl, or H:Cl, which element has the greater electronegativity? _____

b. Because Cl is more electronegative than H, the shared electrons should be _____ (closer to/farther from) the chlorine.

a. yes
b. polar
c. polar

285

Because the shared electrons are closer to the Cl than to the H, the Cl will have a partial negative charge and the H will have a partial positive charge. This may be indicated by a δ (Greek delta) sign.

$$\overset{\delta+ \ \ \delta-}{HCl}$$

Note that this is a partial charge, not an ionic charge.

a. Is one end of the molecule more negative than the other end?

_____

b. The covalent bond between the H and the Cl is _____ (polar/nonpolar).

c. Is the HCl molecule polar or nonpolar? _____

nonpolar

286

Consider the covalent compound XY. If both atoms have the same electronegativity, will the covalent bond between them be polar or nonpolar? _____

287

If there is a difference in electronegativity between X and Y, the covalent bond between them will be _____ and the compound XY will be _____.

288

Now consider the covalent compound CO_2, or O::C::O.

a. Which atom is more electronegative? _____ Which is less electronegative? _____

b. Which atom will have a partial + charge? _____ Which will have a partial − charge? _____

289

The CO_2 molecule may be represented as a linear molecule.

$$\overset{\delta-}{O}::\overset{\delta+}{C}::\overset{\delta-}{O}$$

Both ends of the CO_2 molecule have a partial _____ (positive/negative) charge.

290

a. Does the CO_2 molecule have a positive and a negative end?

_____

b. Is CO_2 a polar or nonpolar compound? _____

nonpolar

291

In general, if a covalent compound is symmetrical, it will be _____ (polar/nonpolar).

H; O

292

In the H_2O molecule, which contains covalent bonds, which atom has a partial positive charge? _____

Which has a partial negative charge? _____

b; a

293

The structure of the water molecule may be represented as either

a. $\overset{\delta+}{H}—\overset{\delta-}{O}—\overset{\delta+}{H}$ or   b. $\overset{\delta+}{H}—\overset{\delta-}{O}$
$\diagdown_{H^{\delta+}}$

The structure in a is theoretical; it does not exist.

Which structure, a or b, represents a polar compound? _____

Which represents a nonpolar compound? _____

b

294

The water molecule has been found to be a polar molecule. Which of the structures in the preceding question more accurately depicts the water molecule? _____

295

One general rule of solubility is that *like dissolves like*. That is, polar liquids dissolve polar compounds, and nonpolar liquids dissolve nonpolar compounds. Is water a polar or a nonpolar liquid? _____

296

a. Should NaCl, a polar compound, dissolve in water? _____

b. Should benzene, a nonpolar compound, dissolve in water? _____

c. Should acetone, a nonpolar liquid, dissolve in benzene, a nonpolar liquid? _____

polar

a. yes
b. no
c. yes

Hydrogen Bonds **10**

297

Consider two water molecules whose structures and partial charges are:

$$\overset{\delta+}{H}\!-\!\overset{\delta-}{O} \qquad \text{and} \qquad \overset{\delta+}{H}\!-\!\overset{\delta-}{O}$$
$$\underset{H^{\delta+}}{\diagdown} \qquad\qquad\qquad \underset{H^{\delta+}}{\diagdown}$$

Which partial charges should attract one another? _____
_____

the negative charge of one water molecule and the positive charge of the other weak

298

Water molecules can form a weak bond between the H of one molecule and the O of another. Such a bond, called a *hydrogen bond*, is caused by the attraction of the partially positive H of one water molecule and the partially negatively charged O (more electronegative) of another water molecule. Hydrogen bonds may be indicated by dotted lines:

Is a hydrogen bond a strong or a weak bond? _____

hydrogen bonds

299

If many water molecules are bonded together in a similar manner, what type of bond should hold them together? _____

300

Hydrogen bonds can also occur between the H attached to an N and an O that is part of a C=O group. Such hydrogen bonds are present in proteins and in DNA.

Note the following diagram, which illustrates a very small part of a protein molecule. (R represents a carbon chain.)

Pleated sheet structure of a protein

What type of bond holds the sections together? _____

hydrogen bonds

301

The hydrogen bonds in the diagram in problem 300 exist between the H attached to a(n) _____ atom and the $=O$ attached to a(n) _____ atom.

N; C

hydrogen bonds

302

We have seen that hydrogen bonding accounts for the pleated sheet structure of some proteins.

Some proteins have a helical structure, as shown in the following diagram:

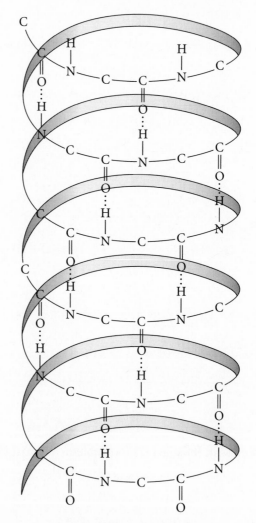

Helical structure of a protein

What type of bond holds the coil in its designated shape? _____

303

In both the pleated sheet and helical structures of proteins, the hydrogen bonds present exist between the H attached to a(n) _____ and the O that is part of a(n) _____ group.

N; C=O

304

Consider the double helical shape of DNA in the following diagram.

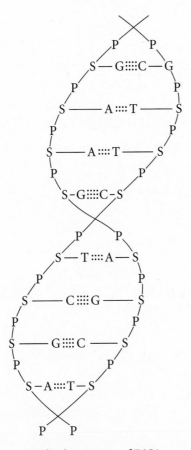

Helical structure of DNA

hydrogen bonds

What type of bond holds the two halves together? _____

2; hydrogen bonds

305

The DNA structure contains the following substances: adenine (A), guanine (G), cytosine (C), and thymine (T).

How many bonds are present between A and T? _____

What type of bonds are these? _____

3; hydrogen bonds

306

How many bonds are present between C and G? _____

What type of bonds are these? _____

weak

307

When the DNA molecule replicates, the two halves come apart, breaking the bonds holding them together.

Are hydrogen bonds strong or weak bonds? _____

easy

308

Should it be easy or difficult to "open up" the DNA chain? _____

Functional Groups in Organic Compounds 11

Functional groups impart certain sets of properties to compounds. One example of such a functional group is the —*OH group*.

ALCOHOLS

309

The —OH group in an ionic compound is called a hydroxide ion.

$$NaOH \longrightarrow Na^+ + OH^-$$

Ionic

The —OH group in a covalent compound is called an *alcohol* group; it does not ionize.

$$\begin{array}{c} H \\ | \\ H-C-OH \\ | \\ H \end{array}$$

Covalent, not ionized

Is a solution of NaOH an electrolyte? _____

yes

no

310

Consider this compound:

$$
\begin{array}{c}
\text{H} \\
| \\
\text{H}-\text{C}-\text{OH} \\
| \\
\text{H}
\end{array}
$$

Is a solution of this compound an electrolyte? _____

The —OH group has the bonds —O—H. It is frequently written as —OH; the bond between the hydrogen and the oxygen is understood and is not written. Note that the oxygen atom still has two bonds attached to it.

In general, covalent compounds containing one or more —OH groups are called *alcohols*.

all of them

311

Which of the compounds a through e is(are) an alcohol? _____

a.
$$
\begin{array}{ccc}
\text{H} & \text{H} & \text{H} \\
| & | & | \\
\text{H}-\text{C}-\text{C}-\text{C}-\text{OH} \\
| & | & | \\
\text{H} & \text{H} & \text{H}
\end{array}
$$

b.
$$
\begin{array}{c}
\text{H} \\
| \\
\text{H}-\text{C}-\text{H} \\
\\
\text{H} \quad\quad \text{H} \\
| \quad\quad | \\
\text{H}-\text{C}-\text{C}-\text{C}-\text{H} \\
| \quad | \quad | \\
\text{H} \quad \text{OH} \quad \text{H}
\end{array}
$$

c. $CH_3CH(OH)CH_3$

d. $CH_3CH_2CH_2CH(CH_3)CH_2OH$

e.
⬡—CH_2CH_2OH
  —CH_3

312

Glycerol, $C_3H_8O_3$, is one of the primary constituents of fats. Its structure is:

$$\begin{array}{ccccc} & H & H & H & \\ & | & | & | & \\ H-&C-&C-&C-&H \\ & | & | & | & \\ & OH & OH & OH & \end{array}$$

Is glycerol an alcohol? _____ Why or why not? _____

yes; it contains
—OH groups

313

The names of alcohols are written either with the word *alcohol* at the end or with the ending *ol*. CH_3OH is methyl alcohol, or methanol. CH_3CH_2OH is ethyl alcohol, or ethanol.

If the name of a compound ends in *ol*, that compound is a(n) _____.

alcohol

314

Which of the following compounds is(are) an alcohol? _____

a.   cholesterol

b.   tocopherol (vitamin E)

c.   thiamine (vitamin B_1)

d.   cortisone

e.   retinol (vitamin A)

a, b, e

ALDEHYDES

A second functional group is the *aldehyde* group. Aldehydes contain a — CHO group. Note that in the aldehyde group the oxygen atom has a double bond attached to it.

$$\underset{}{-\overset{\displaystyle H}{\underset{\displaystyle |}{C}}=O}$$

aldehyde

315

The compound following is an:

$$H-\overset{\displaystyle H}{\underset{\displaystyle H}{C}}-\overset{\displaystyle H}{\underset{\displaystyle H}{C}}-\overset{\displaystyle H}{C}=O$$

__ aldehyde   __ alcohol

The previous compound may also be written as CH_3CH_2CHO, where the — CHO group indicates an aldehyde. Note that the aldehyde group is at the end of the molecule.

Check the correct name for each of the compounds in the next five problems, and explain why you chose the answer you did.

316

```
      H   H   H   H
      |   |   |   |
  H — C — C — C — C — OH
      |   |   |   |
      H   H   H   H
```

__ alcohol   __ aldehyde   __ some other type

Why? _____

alcohol; it has an —OH group

317

```
      H   H
      |   |
  H — C — C = O
      |
      H
```

__ alcohol   __ aldehyde   __ some other type

Why? _____

aldehyde; it has a —CHO group (or equivalent answer)

some other type; it has
neither an —OH group
nor a —CHO group

aldehyde; it has a
—CHO group

alcohol; it has an
—OH group

318

H—C—C—C—H

(structure with H on top of first and third C, H on bottom of first and third C, =O below middle C)

__ alcohol   __ aldehyde   __ some other type

Why? _____

319

__ alcohol   __ aldehyde   __ some other type

Why? _____

320

(benzene ring structure with —C—OH group)

__ alcohol   __ aldehyde   __ some other type

Why? _____

321

Aldehydes are named either with the word *aldehyde* at the end or with the ending *al*.   H   is formaldehyde, or methanal.

$$HC = O$$

CH_3CHO is acetaldehyde, or ethanal.

Retinal is a compound involved in the visual cycle. What kind of compound is it? _____ How do you know? _____

322

a.   The formula for an alcohol group is _____.

b.   The formula for an aldehyde group is _____.

KETONES

Ketones are another functional group. A ketone contains a carbon atom that is attached to a double-bonded oxygen atom and is between two other carbon atoms.

$$-C-C-C-$$
$$\parallel$$
$$O$$

An example of a ketone is acetone, found in the bloodstream and the urine of diabetics.

$$CH_3-C-CH_3$$
$$\parallel$$
$$O$$

Identify the compounds in the next nine problems as alcohols, aldehydes, or ketones.

alcohol

323

$$
\begin{array}{ccccc}
& H & H & H & H & H \\
& | & | & | & | & | \\
H- & C- & C- & C- & C- & C-H \\
& | & | & | & | & | \\
& H & H & H & OH & H
\end{array}
$$

__ alcohol   __ aldehyde   __ ketone

aldehyde

324

$$
\begin{array}{cccc}
& H & H & H & H \\
& | & | & | & | \\
H- & C- & C- & C- & C=O \\
& | & | & | \\
& H & H & H
\end{array}
$$

__ alcohol   __ aldehyde   __ ketone

ketone

325

$$
\begin{array}{ccccc}
& H & H & & H & H \\
& | & | & & | & | \\
H- & C- & C- & C- & C- & C-H \\
& | & | & \| & | & | \\
& H & H & O & H & H
\end{array}
$$

__ alcohol   __ aldehyde   __ ketone

326

$CH_3CH(OH)CH_3$

__ alcohol   __ aldehyde   __ ketone

327

__ alcohol   __ aldehyde   __ ketone

328

__ alcohol   __ aldehyde   __ ketone

alcohol

ketone

ketone

aldehyde and ketone

alcohol and aldehyde

aldehyde and ketone

329

$$CH_3 - \underset{\underset{O}{\|}}{C} - CH_2 - CH_2 - CHO$$

__ alcohol   __ aldehyde   __ ketone

330

$$H - \underset{\underset{H}{|}}{\overset{\overset{H}{|}}{C}} - \underset{\underset{H}{|}}{\overset{\overset{H}{|}}{C}} - \underset{\underset{OH}{|}}{\overset{\overset{H}{|}}{C}} - \underset{\underset{H}{|}}{\overset{\overset{H}{|}}{C}} - \overset{\overset{H}{|}}{C} = O$$

__ alcohol   __ aldehyde   __ ketone

331

$$\underset{\underset{O}{\|}}{\overset{}{\text{—CHO}}}$$
$$\text{—C—CH}_3$$

__ alcohol   __ aldehyde   __ ketone

332

The names of most ketones end in *one*. An example of a ketone is acetone. Which of the following substances are ketones? _____ Alcohols? _____ Aldehydes? _____

a. corticosterone
b. pyridoxal (a B-vitamin)
c. cortisol
d. androsterone
e. estriol
f. pentanal
g. pregnandiol

ORGANIC ACIDS

333

A fourth functional group is the *carboxyl* group, —COOH or

$$-\overset{\overset{\displaystyle O}{\parallel}}{C}-OH$$

The carboxyl group ionizes as follows:

$$-\overset{\overset{\displaystyle O}{\parallel}}{C}-OH \longrightarrow -\overset{\overset{\displaystyle O}{\parallel}}{C}-O^- + H^+$$

Because the carboxyl group yields H^+ ions, it is a(n) _____.

Because organic acids contain a carboxyl group, they are also called *carboxylic acids*.

---

ketones: a, d
alcohols: c, e, g
aldehydes: b, f

acid

a. an aldehyde
b. an acid

334

The —COOH group may also be written as:

$$\underset{\substack{\| \\ O}}{-\text{C}-\text{OH}} \quad \text{or} \quad \underset{\substack{\| \\ \text{OH}}}{-\text{C}=\text{O}}$$

a.   A compound with a —CHO group is called _____.

b.   A compound with a —COOH group is called _____.

carboxyl, or —COOH

335

Citric acid, α-ketoglutaric acid, succinic acid, and oxaloacetic acid are a part of the Krebs cycle, the human body's primary metabolic cycle. What functional group should they all have in common? _____

a. ketone and carboxyl
   (organic acid)
b. alcohol and carboxyl

336

What functional groups are present in the following compounds?

a.   pyruvic acid, produced during the oxidation of glucose _____

$$\text{CH}_3-\underset{\substack{\| \\ O}}{\text{C}}-\text{COOH}$$

b.   lactic acid, formed during the fermentation of milk _____

$$\text{CH}_3-\underset{\substack{| \\ \text{OH}}}{\text{CH}}-\text{COOH}$$

c.   tartaric acid, found in grapes _____

$$
\begin{array}{c}
\text{OH} \\
| \\
\text{H}-\text{C}-\text{COOH} \\
| \\
\text{H}-\text{C}-\text{COOH} \\
| \\
\text{OH}
\end{array}
$$

AMINES

A compound that contains an NH_2 functional group is called an *amine*. It is written as $-NH_2$.

337

Which of the following compounds are amines? _____

a. CH_3NH_2

b. $-CH_2CH_2NH_2$

c. $CH_3CHCH_2CH_3$
         |
       NH_2

d. $CH_2-CH-CH_2-CH_3$
       | |
     NH_2 NH_2

---

c.   alcohol and carboxyl

all of them

a. —COOH (carboxyl)
b. an amine
c. an alcohol
d. an aldehyde
e. a ketone

338

a.   An organic acid is a compound containing a(n) _____ group.

b.   A compound containing an —NH_2 group is called _____.

c.   A compound containing an —OH group is called _____.

d.   A compound containing a —CHO group is called _____.

e.   A compound containing a C—C—C group is called _____.
$$\overset{\|}{\underset{O}{}}$$

Identify the types of compounds in the next five problems. Write out the reasons for your choices.

amine: it contains an —NH_2 group

339

$$H-\overset{\overset{\displaystyle H}{|}}{C}-\overset{\overset{\displaystyle H}{|}}{\underset{\underset{\displaystyle NH_2}{|}}{C}}-\overset{\overset{\displaystyle H}{|}}{\underset{\underset{\displaystyle H}{|}}{C}}-H$$

__ alcohol   __ aldehyde
__ amine   __ acid
__ ketone

Why? _____

340

$$H-\underset{\underset{H}{|}}{\overset{\overset{H}{|}}{C}}-\underset{\underset{OH}{|}}{\overset{\overset{H}{|}}{C}}-\underset{\underset{H}{|}}{\overset{\overset{H}{|}}{C}}-\overset{\overset{H}{|}}{C}=O$$

__ alcohol and acid   __ alcohol and aldehyde
__ amine and alcohol   __ ketone and alcohol

Why? _____

alcohol and aldehyde;
it contains both —CHO
and —OH groups

341

$$H-\underset{\underset{H}{|}}{\overset{\overset{H}{|}}{C}}-\underset{\underset{OH}{|}}{\overset{\overset{H}{|}}{C}}-\underset{\underset{H}{|}}{\overset{\overset{H}{|}}{C}}-\underset{\underset{O}{\|}}{C}-\underset{\underset{H}{|}}{\overset{\overset{H}{|}}{C}}-H$$

__ amine and alcohol   __ alcohol and aldehyde
__ acid and alcohol   __ ketone and alcohol

Why? _____

ketone and alcohol;
it contains both —OH
and C—C—C groups
          ‖
          O

342

$$H-\underset{\underset{H}{|}}{\overset{\overset{H}{|}}{C}}-\underset{\underset{H}{|}}{\overset{\overset{H}{|}}{C}}-\underset{\underset{NH_2}{|}}{\overset{\overset{H}{|}}{C}}-\underset{\underset{H}{|}}{\overset{\overset{H}{|}}{C}}-\underset{\underset{H}{|}}{\overset{\overset{H}{|}}{C}}-OH$$

Type _____

Why? _____

amine and alcohol,
or amino alcohol; it
contains both —NH$_2$
and —OH groups

amine and organic acid;
it contains both —NH$_2$
and —COOH groups

343

$$H-\overset{\overset{\displaystyle H}{|}}{\underset{\underset{\displaystyle H}{|}}{C}}-\overset{\overset{\displaystyle H}{|}}{\underset{\underset{\displaystyle NH_2}{|}}{C}}-\overset{\overset{\displaystyle H}{|}}{\underset{\underset{\displaystyle H}{|}}{C}}-\overset{\overset{\displaystyle O}{||}}{C}-OH$$

Type _____

Why? _____

Identify the types of compounds in the next four problems.

alcohol and organic acid

344

COOH

OH

Type _____

amine and acid
(amino acid)

345

CH$_3$CH(NH$_2$)COOH

Type _____

$$\boxed{346}$$

$$\underset{\underset{O}{\overset{\|}{}}}{CH_3CCH_2CH_2NH_2}$$

Type _____

ketone and amine, or aminoketone

$$\boxed{347}$$

$$CH_3CH_2CH(OH)COOH$$

Type _____

alcohol and acid

ESTERS

A compound that contains a $\underset{\underset{}{\overset{O}{\overset{\|}{}}}}{-C-O-C}$ group is called an *ester*. The following compound is an ester because it contains such a group:

$$\underset{\underset{H}{|}}{\overset{\overset{H \quad\quad O \quad\quad H}{}}{H-C-C-O-C-H}}\underset{H}{|}$$

$$\boxed{348}$$

Fats and oils are esters. They contain a _____ group.

$\underset{\underset{}{\overset{O}{\overset{\|}{}}}}{-C-O-C}$

yes

349

$$\text{C}_6\text{H}_5 - \overset{\overset{\displaystyle O}{\|}}{\text{C}} - \text{O} - \text{CH}_3$$

Is this compound an ester? _____

AMIDES

A compound that contains a $-\overset{\overset{\displaystyle O}{\|}}{\text{C}} - \text{NH}$ or $-\overset{\overset{\displaystyle O}{\|}}{\text{C}} - \text{NH}_2$ group is called an *amide*.

The following compound, niacinamide, which is derived from the B-vitamin niacin, is an example of an amide.

Niacinamide

350

Cells link amino acids together. The bond holding amino acids together is called an *amide bond* or a *peptide bond*.

A dipeptide is formed by the linkage of how many amino acids?

_____

The amino acids in a dipeptide are held together by an amide or a(n) _____ bond.

A tripeptide consists of _____ (how many?) amino acids held together by _____ or _____ bonds.

2; peptide; 3; amide; peptide

SULFHYDRYLS

A compound that contains an —SH group is called a *sulfhydryl* or a *thiol*.

The following compound, cysteine, one of the essential amino acids, is a sulfhydryl, or thiol, because it contains an —SH group.

$$NH_2 - \underset{\underset{CH_2 - SH}{|}}{\overset{\overset{COOH}{|}}{C}} - H$$

Cysteine

351

The activity of some enzymes is due to the presence of sulfhydryl groups. Cysteine also contains which other functional groups?

_____

acid (or —COOH) and —NH$_2$ (or amine)

alcohol; sulfur; hydrogen

352

The prefix *thio* refers to the element sulfur, whereas the suffix *ol* indicates a(n) _____. So, a thiol is a sulfur alcohol and is made from the elements _____ and _____.

DISULFIDES

Organic disulfides or disulfide bridges contain an —S—S— group.

The following compound, cystine, contains a disulfide bridge.

$$
\begin{array}{ccc}
\text{COOH} & & \text{COOH} \\
| & & | \\
\text{NH}_2\text{—CH} & \text{NH}_2\text{—CH} \\
| & & | \\
\text{CH}_2\text{—S—S—CH}_2
\end{array}
$$

Cystine

2; sulfur; 2; sulfur

353

The prefix *di* indicates the number _____. Sulfide refers to the element _____. So, a disulfide bridge contains _____ (how many?) atoms of the element _____.

354

Note the following abbreviated structure of human insulin.

What type of bonds or linkages holds the A and B chains together?

_____

How many bonds of this type are present in human insulin? _____

PHOSPHATES

A compound that contains an —OPO_3H_2 group is called a *phosphate*. This group may also be written as

$$-O-\overset{\displaystyle \overset{O}{\|}}{\underset{\displaystyle \underset{OH}{|}}{P}}-OH$$

The following compound is a phosphate because it contains the —OPO_3H_2 group:

$$H-\overset{\displaystyle \overset{H}{|}}{\underset{\displaystyle \underset{H}{|}}{C}}-\overset{\displaystyle \overset{O}{\|}}{C}-O-\overset{\displaystyle \overset{O}{\|}}{\underset{\displaystyle \underset{OH}{|}}{P}}-OH$$

355

The following structure represents adenosine monophosphate (AMP). Draw a circle around the phosphate group. This compound also contains an amine and two alcohol groups. Draw a square around the amine group, and a rectangle around the OH groups.

SUMMARY OF FUNCTIONAL GROUPS

Name the groups present in the compounds in problems 356–368.

$$\boxed{356}$$

$$
\begin{array}{c}
\text{CHO} \\
| \\
\text{H} - \text{C} - \text{OH} \\
| \\
\text{HO} - \text{C} - \text{H} \\
| \\
\text{H} - \text{C} - \text{OH} \\
| \\
\text{H} - \text{C} - \text{OH} \\
| \\
\text{H} - \text{C} - \text{OH} \\
| \\
\text{H}
\end{array}
$$

Glucose

aldehyde and alcohols (more than one alcohol group)

$$\boxed{357}$$

$$CH_3CH(NH_2)COOH$$

Alanine

amine and acid (amino acid)

$$\boxed{358}$$

Methyl salicylate

alcohol and ester

aldehyde and alcohols

359

Pyridoxal
(one of the B-vitamins)

one phosphate group
and two ester groups

360

Lipid

two alcohol groups, a
ketone group, and an
acid group

361

Prostaglandin

362

$$\begin{array}{c} COOH \\ | \\ CHNH_2 \\ | \\ CH_2OH \end{array}$$

Serine

an alcohol group, an
amine group, and an
acid group

363

$$\begin{array}{c} CHO \\ | \\ HC-OH \\ | \\ H_2C-OPO_3H_2 \end{array}$$

Glyceraldehyde-3-P

aldehyde, alcohol, and
phosphate groups

364

$$\begin{array}{c} O \\ \| \\ CH_2-OC-C_{17}H_{35} \\ | \qquad\quad O \\ \qquad\quad \| \\ CH-O-C-C_{17}H_{35} \\ | \qquad\quad O \\ \qquad\quad \| \\ CH_2-OP-OCH_2CH_2NH_2 \\ | \\ OH \end{array}$$

Cephalin

two ester groups, a
phosphate group, and
an amine group

an amide group

365

$$O$$
$$\parallel$$
C—NH$_2$

(benzene ring attached to C)

a thiol (or sulfhydryl) group

366

CH$_3$CH$_2$CH$_2$SH

two acid groups, two amine groups, and one disulfide bridge

367

COOH   COOH
| |
NH$_2$CH   NH$_2$CH
| |
H$_2$C — S — S — CH$_2$

an acid group, an amine group, and a sulfhydryl group

368

COOH
|
NH$_2$CH
|
CH$_2$ — SH

369

The following diagram illustrates part of the shape of a protein molecule.

What type of bond or bridge determines the conformation of such a protein? _____

370

Polypeptides are made from many amino acids bonded together by _____ or _____ bonds.

disulfide

amide; peptide

PART **III**

Biochemistry

Carbohydrates 12

Carbohydrates contain the elements carbon, hydrogen, and oxygen, and only those elements. A compound containing elements other than these three is not a carbohydrate.

The word *hydrate* refers to water. Many carbohydrates contain hydrogen and oxygen in the same ratio as in water—2:1.

371

Which of the following compounds are carbohydrates?

___ C_6H_6 ___ $C_{12}H_{22}O_{11}$ ___ $C_5H_{10}O_5$

___ $C_6H_{12}O_6N$ ___ C_2H_6O

$C_{12}H_{22}O_{11}$ and $C_5H_{10}O_5$

372

a. In the list of compounds in problem 371, why is $C_6H_{12}O_6N$ not a carbohydrate? _____

_____

b. Why are C_6H_6 and C_2H_6O not carbohydrates? _____

_____

a. $C_6H_{12}O_6N$ contains nitrogen, and carbohydrates contain only C, H, and O

b. C_6H_6 contains no oxygen, and C_2H_6O has the wrong ratio of H to O

a. one
b. one

all of them

5

373

Carbohydrates are divided into three types. One type of carbohydrate is the *monosaccharide*.

a. The word *saccharide* means *simple sugar*. The prefix *mono* means
_____.

b. Therefore, monosaccharides are simple sugars, each containing only
_____ simple sugar.

374

Monosaccharides and disaccharides have names ending in *ose*. Which of the following are monosaccharides or disaccharides?

__ glucose   __ fructose   __ galactose
__ arabinose   __ erythrose

375

Monosaccharides are named according to the number of carbon atoms they contain. A pentose contains _____ carbon atoms.

An example of a pentose is ribose, $C_5H_{10}O_5$. The structure of ribose is:

In this type of structure the ring, consisting of carbon atoms at each corner and an oxygen atom where indicated, is considered to be in a plane perpendicular to the paper. The heavy line in the ring indicates the section closest to you; the light lines, the part farther from you. The —H and —OH groups are above and below the plane as indicated.

Ribose occurs in ribonucleic acid, RNA, which will be discussed later.

376

Compare the structures of ribose, $C_5H_{10}O_5$, and deoxyribose, $C_5H_{10}O_4$.

Ribose   Deoxyribose

a.   The prefix *deoxy* means _____.

b.   Deoxyribose contains _____ than does ribose.

377

Ribose occurs in ribonucleic acid, RNA. Deoxyribose occurs in deoxyribonucleic acid, _____.

---

a. one less oxygen
b. one less oxygen

DNA

a. 5
b. 6

ose

378

a. A pentose is a monosaccharide containing _____ carbon atoms.

b. A hexose is a monosaccharide containing _____ carbon atoms.

A primary hexose in the body is glucose, $C_6H_{12}O_6$. The oxidation of glucose supplies much of the energy the body needs. The structure of glucose is:

379

The second type of carbohydrate is the *disaccharide*. Both monosaccharides and disaccharides have names ending in _____.

380

The prefix *di* means *two*. Disaccharides are formed by the combination of two monosaccharides, with water also being produced. An enzyme is required for this reaction.

monosaccharide + monosaccharide $\xrightarrow{\text{enzyme}}$ disaccharide + water

Conversely, disaccharides react with water to produce two monosaccharides.

disaccharide + water $\xrightarrow{\text{enzyme}}$ monosaccharide + _____

These reactions may be written as shown on the following page.

The bond holding the two halves of the disaccharide together is called a *glycosidic bond*, or a *glycosidic linkage*.

monosaccharide

381

When a disaccharide reacts with water, two monosaccharides are produced. Such a reaction is called *hydrolysis*, which is defined as the breaking apart of a molecule by reaction with water.

Digestion involves hydrolysis. During digestion, molecules are

___ broken down into smaller molecules
___ built up into larger molecules

broken down into
smaller molecules

monosaccharides

382

Disaccharides that have undergone hydrolysis yield two _____.

Glucose
(a monosaccharide)

+

Glucose
(a monosaccharide)

Enzyme   Enzyme
(hydrolysis)   (synthesis)

Maltose
(a disaccharide)

+ H_2O

Water

one; two

383

Monosaccharides are made up of _____ simple sugar(s).
Disaccharides are made up of _____ simple sugar(s).

384

The hydrolysis of a disaccharide may be written as:

$$C_{12}H_{22}O_{11} \ + \ H_2O \ \xrightarrow{\text{enzyme}} \ C_6H_{12}O_6 \ + \ C_6H_{12}O_6$$

Disaccharide + water   Monosaccharide + monosaccharide

Sucrose, $C_{12}H_{22}O_{11}$, is a disaccharide. The hydrolysis of sucrose yields the two _____, glucose and fructose.

> monosaccharides

385

There are three common disaccharides: *sucrose*, *maltose*, and *lactose*. These disaccharides all have the same molecular formula, $C_{12}H_{22}O_{11}$. Lactose, also called milk sugar, is found in milk. Maltose, or malt sugar, is found in sprouting grain. Sucrose, or cane sugar, is found in _____.

> sugarcane

386

The third type of carbohydrate is the *polysaccharide*. The prefix *poly* means *many*, so polysaccharides, on hydrolysis, yield many _____.

> monosaccharides

387

Recall that hydrolysis means reaction with _____ to produce _____ (smaller/larger) molecules.

> water; smaller

$C_{12}H_{22}O_{11}$

388

Starch is a polysaccharide. Its molecular formula is $(C_6H_{10}O_5)_n$, where n is some large number. Upon complete hydrolysis, starch yields many monosaccharides:

$$(C_6H_{10}O_5)_n \; + \; nH_2O \; \xrightarrow{\text{enzyme}} \; nC_6H_{12}O_6$$

Starch,
a polysaccharide   Monosaccharides

Which of the following formulas represents a disaccharide?

__ $C_5H_{10}O_5$ __ $C_{12}H_{22}O_{11}$

__ $(C_6H_{10}O_5)_n$ __ $C_6H_{12}O_6$

$(C_6H_{10}O_5)_n$; $C_5H_{10}O_5$ and $C_6H_{12}O_6$

389

Which of the formulas listed in problem 388 represents a poly-saccharide? _____

Which are monosaccharides? _____

a. glucose, fructose, galactose
b. lactose, maltose, sucrose
c. starch, cellulose, glycogen

390

Other examples of polysaccharides are *cellulose*, found in plants, and *glycogen*, found in animals. Plants use the polysaccharide cellulose primarily for support. Animals store carbohydrate in the form of a polysaccharide, glycogen.

a. Name three monosaccharides: _____

b. Name three common disaccharides: _____

c. Name three polysaccharides: _____

391

Disaccharides are held together by a glycosidic bond. Polysaccharides are held together by _____ bonds.

many glycosidic

392

a. Hydrolysis of a disaccharide yields _____.

b. Hydrolysis of a polysaccharide yields _____.

a. two monosaccharides
b. many monosaccha-
rides

PROPERTIES OF CARBOHYDRATES

Monosaccharides are crystalline compounds, soluble in water and sweet to the taste, and need no digestion in order to be absorbed into the bloodstream.

Disaccharides are crystalline, water-soluble, and sweet to the taste and must be hydrolyzed (digested) to monosaccharides before they can be absorbed and used by the body for energy.

Polysaccharides are not water-soluble and are not crystalline. They form colloidal dispersions instead of solutions. They are not sweet and must be hydrolyzed before being absorbed. (Cellulose is not hydrolyzed by enzymes in the human gastrointestinal tract.)

a. mono- and disaccharides
b. polysaccharides
c. mono- and disaccharides
d. polysaccharides
e. mono- and disaccharides
f. polysaccharides
g. monosaccharides

393

a. Which type(s) of carbohydrate is(are) soluble in water? _____

b. Which type(s) of carbohydrate form(s) a colloidal suspension in water? _____

c. Which type(s) of carbohydrate is(are) crystalline? _____

d. Which is(are) not crystalline? _____

e. Which is(are) sweet? _____

f. Which is(are) not sweet? _____

g. Which require(s) no digestion? _____

Lipids 13

Lipids include such substances as fats, phospholipids, and steroids. Lipids are insoluble in water. Lipids contain the same elements as carbohydrates, but in lipids the proportion of oxygen is much lower than in carbohydrates.

| 394 |

Lipids contain the elements _____, _____, and _____.

carbon; hydrogen; oxygen

FATS

| 395 |

Fats are esters. Esters are formed by the reaction of an organic acid with an alcohol. In fats, the organic acid is called a *fatty acid* and the alcohol is *glycerol* (sometimes called glycerin). The reaction is:

$$\text{fatty acids} + \text{glycerol} \xrightarrow{\text{enzymes}} \text{fat} + \text{water}$$

The reverse reaction is the hydrolysis of a fat.

Thus, the hydrolysis of a fat yields _____ + _____.

fatty acids; glycerol

fat

fatty acid; glycerol; esters

396

Fatty acids and glycerol are products of the hydrolysis of a _____.

397

The chemical equation for the hydrolysis of a fat may be written as

$$
\begin{array}{c}
\underset{\text{A}}{\underbrace{
\begin{array}{l}
\ \ \ \ \ \ \ \overset{\displaystyle H}{\underset{\displaystyle |}{}} \ \ \ \ \ \ \overset{\displaystyle O}{\underset{\displaystyle \|}{}} \\
H-C-O-C-C_{17}H_{35} \\
\ \ \ \ \ \ \ | \ \ \ \ \ \ \ \ \ \overset{\displaystyle O}{\underset{\displaystyle \|}{}} \\
H-C-O-C-C_{17}H_{35} \\
\ \ \ \ \ \ \ | \ \ \ \ \ \ \ \ \ \overset{\displaystyle O}{\underset{\displaystyle \|}{}} \\
H-C-O-C-C_{17}H_{35} \\
\ \ \ \ \ \ \ | \\
\ \ \ \ \ \ \ H
\end{array}
}} + 3H_2O \longrightarrow
\underset{\text{B}}{\underbrace{3C_{17}H_{35}COOH}} +
\underset{\text{C}}{\underbrace{
\begin{array}{l}
\ \ \ \ \ \ \ H \\
\ \ \ \ \ \ \ | \\
H-C-OH \\
\ \ \ \ \ \ \ | \\
H-C-OH \\
\ \ \ \ \ \ \ | \\
H-C-OH \\
\ \ \ \ \ \ \ | \\
\ \ \ \ \ \ \ H
\end{array}
}}
\end{array}
$$

Substance A is a fat.

Substance B is a _____.

Substance C is an alcohol called _____.

Substance A contains what type of groups? _____

398

a.   How do you know that substance B is an acid? _____

b.   How do you know that substance C is an alcohol? _____

c.   Why is substance A called an ester? _____

a.   It contains —COOH groups.
b.   It contains —OH groups.
c.   It contains

$$\overset{O}{\overset{\|}{C}}-O-C\left(\overset{O}{\overset{\|}{C}}-O-C\right)$$

groups.

399

a.   The hydrolysis of a polysaccharide yields _____.

b.   The hydrolysis of a disaccharide yields _____.

c.   The hydrolysis of a fat yields _____.

a.   many monosaccharides
b.   two monosaccharides
c.   fatty acids and glycerol

400

Fatty acids that contain double bonds are called *unsaturated*. Unsaturated fatty acids tend to be more liquid than saturated fatty acids. A liquid fat is called an *oil*.

Increasing the number of double bonds between carbon atoms in fatty acids of a fat molecule should make the fat more _____ (solid/liquid). Decreasing the number of double bonds in a fat will make the fat more _____ (solid/liquid).

liquid; solid

401

Which are unsaturated, fats or oils? _____

oils

those containing many
double bonds

402

Unsaturated compounds contain double bonds. What kind of compounds would be called polyunsaturated? _____

solid

403

Three fatty acids, butyric (C_3H_7COOH), palmitic ($C_{15}H_{31}COOH$), and stearic ($C_{17}H_{35}COOH$) acid, contain only single bonds. When they are present in a lipid, will that lipid tend to become solid or liquid? _____

some double bonds

404

Arachidonic acid is unsaturated. This means that it contains _____ (only single bonds/some double bonds) in its structure.

an oil; the fatty acid
is unsaturated

405

A lipid that undergoes hydrolysis yields arachidonic acid and glycerol. Was the lipid a fat or an oil? _____

How do you know? _____

nonpolar

406

Lipids are insoluble in water. They are soluble in organic solvents such as gasoline, ether, carbon tetrachloride, and benzene. Organic solvents are nonpolar. Recall that in general, polar compounds dissolve in polar liquids, and nonpolar compounds dissolve in nonpolar liquids. Are lipids polar or nonpolar? _____

PHOSPHOLIPIDS

407

The hydrolysis of a phospholipid may be written as:

$$\text{phospholipid} \xrightarrow[\text{enzymes}]{\text{hydrolysis}} \text{fatty acids} + \text{glycerol} + \text{phosphoric acid} + \text{an N-compound}$$

The hydrolysis of a fat yields _____ + _____.

> fatty acids; glycerol

408

The N-compounds in phospholipids are a certain group of nitrogen-containing compounds. Among them are serine, choline, inositol, and ethanolamine. Which is present determines the type of phospholipid, as will be discussed in the following problems.

The hydrolysis of a phospholipid yields _____

_____.

> fatty acids, glycerol, phosphoric acid, and a nitrogen compound

409

a. Lipids contain which elements? _____

b. Phospholipids contain which elements? _____

> a. C, H, O
> b. C, H, O, P, N

410

Lecithin is a phospholipid in which the N-compound is *choline*. Lecithin is a main component of cell membranes and is involved in the transportation of fats in the body. The hydrolysis of lecithin yields fatty acids, glycerol, phosphoric acid, and _____.

> choline

fatty acids; glycerol; phosphoric acid; ethanolamine

$$\boxed{411}$$

Cephalin is a phospholipid involved in the blood-clotting process. In cephalin the N-compound is *ethanolamine*.

cephalin $\xrightarrow[\text{enzymes}]{\text{hydrolysis}}$ _____, _____,

_____, and _____.

STEROIDS

a. insoluble
b. insoluble
c. soluble

$$\boxed{412}$$

Another group, called steroids, is classified along with lipids because of similar solubility characteristics.

a.   Lipids are _____ (soluble/insoluble) in water.

b.   Steroids are _____ (soluble/insoluble) in water.

c.   Steroids are _____ (soluble/insoluble) in organic liquids.

a. no
b. no

$$\boxed{413}$$

Steroids are high-molecular-weight tetracyclic (4-ring) compounds whose structure is similar to this:

a.   Do steroids contain fatty acids? _____

b.   Do steroids contain glycerol? _____

414

Examples of steroids in the body are cholesterol, bile salts, sex hormones, and hormones of the adrenal cortex.

a. Does cholesterol have a steroid structure? _____

b. The ending *ol* in cholesterol indicates that it is what type of compound? _____

a. yes
b. an alcohol

Proteins 14

FORMATION AND HYDROLYSIS

415

Proteins contain the same elements as do carbohydrates and fats, except that proteins always contain one additional element: nitrogen.

All proteins contain the four elements: _____.

> carbon, hydrogen, oxygen, nitrogen (in any order)

416

Some proteins also contain additional elements, such as sulfur, phosphorus, or iron. The hydrolysis of proteins yields amino acids (see problem 417). Likewise, the combination of amino acids yields proteins.

$$\text{protein} \underset{\text{combination}}{\overset{\text{hydrolysis}}{\rightleftarrows}} \text{amino acids}$$

Amino acids contain what functional groups? _____

> amine and organic acid, or carboxyl (or —NH_2 and —COOH)

— COOH or carboxyl

417

The structure of an amino acid can be represented as

$$\underset{H}{\overset{H}{\diagdown}} N - \underset{R}{\overset{H}{\underset{|}{C}}} - \underset{}{\overset{O}{\overset{\parallel}{C}}} - OH \quad or \quad H_2N - \underset{R}{\overset{}{\underset{|}{CH}}} - COOH$$

where the "R" represents either a hydrogen or carbon-hydrogen groups.

Amino acids (and also proteins) can act as buffers because the —NH_2 group acts as a H^+ acceptor and the _____ group acts as a H^+ donor.

a. 3
b. many (or more than 3)

418

When two amino acids combine, the product is called a *dipeptide*.

a.   A *tripeptide* would be formed when _____ amino acids combine.

b.   A *polypeptide* would be formed when _____ amino acids combine.

amino acids

419

Note the formulas for alanine and glycine:

$$\underset{NH_2}{\overset{CH_3CHCOOH}{\underset{|}{}}} \qquad \underset{NH_2}{\overset{CH_2COOH}{\underset{|}{}}}$$

         Alanine   Glycine

Both would be classified as what type of compound? _____

420

When these two amino acids react, the following reactions are possible:

| Alanine | Glycine | Alanyl-glycine (ala-gly) |

| Glycine | Alanine | Glycyl-alanine (gly-ala) |

Note that in both of these equations, the reaction is between the OH of a(n) _____ group and the H of a(n) _____ group.

organic acid (or carboxyl); amine

421

When glycine and alanine react, what type of compound is formed?

_____

a dipeptide

422

a. The abbreviation *ala* represents _____.

b. The abbreviation *ala-gly* represents _____.

a. alanine
b. alanyl-glycine

a tripeptide

423

If the amino acids glycine, alanine, and valine were combined in one compound, what type would it be? _____

ala-gly-val

424

What would be the abbreviation for a tripeptide containing alanine, glycine, and valine, in that order? _____

a tripeptide containing glycine, valine, and alanine, combined in that order

425

What would the abbreviation *gly-val-ala* represent? _____
_____

many; peptide

426

A dipeptide consists of two amino acids held together by a *peptide bond*. A tripeptide consists of three amino acids held together by peptide bonds. A polypeptide consists of _____ amino acids held together by _____ bonds.

amide

427

Peptide bonds are also called _____ bonds. (See problem 350.)

428

In a peptide bond (see problem 420), the bond occurs between a
_____ atom of one amino acid and a _____ atom
of another amino acid.

STRUCTURE

The *primary structure* of a protein refers to the linear sequence of amino
acids in that protein. An example of a primary structure is:

$$NH_2-CH_2-\overset{\overset{\displaystyle O}{\|}}{C}-NH-\underset{\underset{\displaystyle CH_3}{|}}{CH}-\overset{\overset{\displaystyle O}{\|}}{C}-NH-\underset{\underset{\underset{\displaystyle CH_3}{|}}{\underset{\displaystyle CH_2}{|}}}{CH}-\overset{\overset{\displaystyle O}{\|}}{C}-OH$$

Amino acid   Amino acid   Amino acid

429

What type of bond holds the amino acids together in the primary
structure? _____

Note that all proteins have an amino ($-NH_2$) terminus and a carboxyl
($-COOH$) terminus.

The *secondary structure* of a protein refers to the regular recurring arrangement of the amino acid chain (its primary structure). One such arrangement, called the α-helix, occurs when the amino acid chain forms a spiral or coil, as shown here:

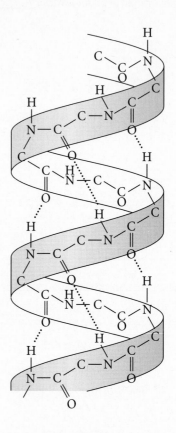

430

hydrogen

The coil consists of loops of the linear arrangement of the amino acids. These loops are held together by bonds between the O of the C=O of one amino acid and the H of the NH of another amino acid.

Such a bond is called a _____ bond (refer to problems 297–308 if you don't know).

431

Another type of secondary structure, called the β-pleated sheet, consists of parallel strands of polypeptides as shown in the following figure:

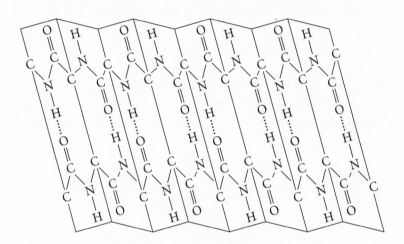

What type of bonds holds these sheets together? _____

hydrogen

432

a. The linear sequence of amino acids in a protein is called the _____ structure of that protein.

b. The regular recurring arrangement of the amino acid chains is called the _____ structure of a protein.

a. primary
b. secondary

a. *α*-helix; *β*-pleated
   sheet
b. hydrogen

433

a. Two types of secondary structures are possible. They are the
    _____ and the _____.

b. What type of bond holds these secondary structures together?
    _____

The *tertiary structure* of a protein refers to the specific folding and bending of the coils into layers or fibers, as shown here. It is this tertiary structure that gives proteins (and enzymes, which are also proteins) their biological activity.

Some proteins have a *quaternary structure* that occurs when several protein units, each with its own primary, secondary, and tertiary structure, combine to form a more complex unit, as shown in Figure **e**. An example of a protein with a quaternary structure is hemoglobin.

a

c

b   d   e

434

Look at the preceding five structures and label each as to primary, secondary, tertiary, or quaternary structure.

Which structure represents an α-helix? _____

Which structure represents a β-pleated sheet? _____

What type of bond holds the primary structures together? _____

What type of bond holds the secondary structure together? _____

a = primary;
b = secondary;
c = secondary;
d = tertiary;
e = quaternary

b; c; peptide; hydrogen

DENATURATION

weak

435

Hydrogen bonds are _____ (strong/weak) bonds.

no

436

Therefore, hydrogen bonds are easily broken. If the hydrogen bonds in a protein are broken, can the protein easily maintain its structural shape? _____

cannot

437

If the hydrogen bonds in a protein are broken, the shape of the protein can change, making it incapable of performing its physiologic function. The protein is said to be *denatured*.

The activity of many proteins (and enzymes) depends upon the presence of an active site that "fits" into a specific substrate. In Figure **a** following, the active site is shown in the center of the structure.

a   b

Enzymes

If the protein is denatured, the parts of the active site are no longer in close proximity (Figure **b**), can no longer "fit" into the substrate, and therefore _____ (can/cannot) react.

438

If the hydrogen bonds in a protein can easily re-form, then the denaturation is said to be *reversible*. If the hydrogen bonds cannot re-form, then the process is termed *irreversible denaturation*.

a. In the denaturation of protein, _____ bonds are broken.

b. If the broken hydrogen bonds can easily re-form, the process is called _____ denaturation.

c. If the broken hydrogen bonds cannot re-form, the process is called _____.

Factors that can denature proteins include strong acids, heat, alcohol, and salts of heavy metals.

a. hydrogen
b. reversible
c. irreversible denaturation

Nucleotides 15

ADENOSINE TRIPHOSPHATE

439

Nucleotides are formed by the reaction of a pentose (labeled Ⓢ for sugar), phosphoric acid (labeled Ⓟ), and a nitrogen-containing base (labeled Ⓝ), as shown:

Ⓝ — Ⓢ — Ⓟ

One example of a nucleotide is adenosine monophosphate, AMP, which is formed by the reaction of adenine (a nitrogen-containing base), ribose (a pentose), and phosphoric acid. From the name *adenosine monophosphate*, how many phosphoric acid molecules are present? _____

440

The structure of adenosine monophosphate, AMP, is shown in the following figure. It can be abbreviated as adenosine-P.

1

ribose; adenine

AMP

AMP contains which pentose? _____

Which nitrogen-containing base? _____

3

441

If a second molecule of phosphoric acid is bonded to the first
phosphoric acid group, adenosine diphosphate, ADP, is formed.
In adenosine triphosphate, ATP, how many phosphoric acid groups
are present? _____

energy

442

ATP provides a form of chemical energy that is usable by all body cells.
The structure of ATP may be abbreviated as

$$\text{adenosine} - \text{P} \sim \text{P} \sim \text{P}$$

Note that ATP contains two high-energy phosphate bonds indicated
as $\sim$.

In the reaction

$$\text{ATP} \longrightarrow \text{ADP} + \text{P} + \text{?} \qquad \text{or}$$

$$\text{adenosine} - \text{P} \sim \text{P} \sim \text{P} \longrightarrow \text{adenosine} - \text{P} \sim \text{P} + \text{P} + \text{?}$$

what is produced other than ADP and P? _____

443

When ADP is converted into AMP, is energy released? _____

yes

444

To convert AMP back to ATP, what must be added? _____

energy and two phosphoric acid groups

445

Much of the energy required to form ATP comes from the metabolism of glucose.

a. AMP contains _____ phosphoric acid group(s) and _____ high-energy phosphate bond(s).

b. ADP contains _____ phosphoric acid group(s) and _____ high-energy phosphate bond(s).

c. ATP contains _____ phosphoric acid group(s) and _____ high-energy phosphate bond(s).

a. 1; 0
b. 2; 1
c. 3; 2

446

If two nucleotides are joined together, a dinucleotide is formed. One such dinucleotide is nicotinamide adenine dinucleotide, NAD.

a. Both nucleotides that make up NAD must contain which pentose? _____

b. One nucleotide contains nicotinamide as the nitrogen-containing base (N-base); the other contains _____ as the nitrogen-containing base.

a. ribose
b. adenine

flavin, adenine

447

Which nitrogen-containing bases are present in flavin adenine dinucleotide, FAD? _____

Both dinucleotides, FAD and NAD, are involved in oxidation-reduction reactions in cells.

NUCLEIC ACIDS: DNA AND RNA

Nucleic acids are polymers of nucleotides. Consider the following structure of deoxyribonucleic acid, DNA, a nucleic acid. Note that each boxed segment is a nucleotide.

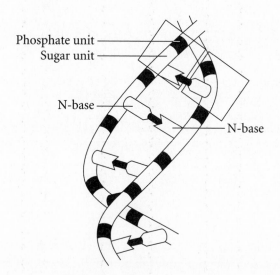

Phosphate unit
Sugar unit

N-base

N-base

DNA consists of a double coil with the opposite sides held together by hydrogen bonds. The backbones of the coils are alternating sugar and phosphate groups, with the hydrogen bonding between the nitrogen-containing bases holding the sides together.

448

In deoxyribonucleic acid, which pentose is a part of the coil backbone?

_____

deoxyribose

449

In DNA there are four N-bases: adenine, guanine, thymine, and cytosine. They are abbreviated by the first letter (capitalized) of their name.

Adenine is abbreviated as _____.

Guanine is abbreviated as _____.

Thymine is abbreviated as _____.

Cytosine is abbreviated as _____.

A; G; T; C

450

Which N-base is represented by each of the following abbreviations?

T _____

A _____

C _____

G _____

thymine; adenine; cytosine; guanine

2; 3

In DNA adenine is always hydrogen-bonded to thymine, and cytosine is always hydrogen-bonded to guanine. Thus the structure of DNA may be represented as:

How many hydrogen bonds hold A and T together? _____

How many hydrogen bonds hold C and G together? _____

452

The four N-bases in DNA are: _____

_____.

adenine, A; thymine, T; guanine, G; and cytosine, C

453

In DNA, adenine is always hydrogen-bonded to _____, and guanine is always bonded to _____.

454

Complete the structure of the following figure, indicating missing N-bases and hydrogen bonds.

455

A second nucleic acid is ribonucleic acid, RNA. In RNA the pentose is ribose; in DNA the pentose is _____.

456

In RNA, A is always bonded to uracil, U. In DNA, A is always bonded to _____.

a. G; 3
b. G; 3
c. T; 2
d. U; 2
e. deoxyribose
f. ribose

457

a. In DNA, C bonds to _____ with _____
    (how many?) hydrogen bonds.

b. In RNA, C bonds to _____ with _____
    (how many?) hydrogen bonds.

c. In DNA, A bonds to _____ with _____
    (how many?) hydrogen bonds.

d. In RNA, A bonds to _____ with _____
    (how many?) hydrogen bonds.

e. In DNA the sugar is _____.

f. In RNA the sugar is _____.

REPLICATION OF DNA

458

Consider the following illustration in which a DNA segment has been
straightened out and the S's (sugars) and P's (phosphates) understood
and not printed.

If the DNA strands are separated, then each strand can act as a template
to attract additional N-bases.

Original
strand
opened up

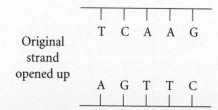

Let us consider the T and A at the far left of the open strand. The T will attract an A and the A will attract a T to begin the formation of a new backbone of a DNA molecule, as follows.

Now complete the balance of the strands for each DNA molecule, again showing N-bases and hydrogen bonding.

459

Are the two new DNA strands the same as the original one?

_____

This is the basis of DNA replication, which results in two new copies that are identical to the original strand.

TRANSCRIPTION OF INFORMATION (PROTEIN SYNTHESIS)

460

During transcription, DNA "unzips" and only part of one segment acts as a template for the formation of messenger RNA (mRNA).

In DNA, the partner of adenine (A) is _____.

In mRNA, the partner of adenine (A) is uracil, or _____.

yes

T, or thymine; U

2; U; 2

In DNA, A is bonded to T with how many hydrogen bonds?

_____

Likewise in mRNA, A is bonded to _____ with how many hydrogen bonds? _____

deoxyribose; ribose

461

In DNA the sugar is _____.

In mRNA the sugar is _____.

462

Complete the following diagram indicating the partners and bonds needed for the formation of mRNA from the indicated DNA segment.

$$\begin{array}{c} \text{Segment} \\ \text{of DNA} \end{array} \quad \overline{}$$
$$\qquad\qquad\qquad C \quad A \quad G \quad T \quad A \quad C$$

mRNA   _____

mRNA acts as a template for the transmission of genetic information from the cell's genes to the site of protein synthesis.

In the cytoplasm another type of RNA is present. It is called transfer RNA or tRNA. The function of tRNA is to direct the production of the protein called for by mRNA.

tRNA has three N-bases at one end and an amino acid at the other end. It may be indicated by the following structure:

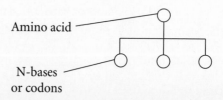

Amino acid

N-bases
or codons

As mRNA travels across a ribosome, the first group of three N-containing bases picks up the corresponding tRNA. As the mRNA continues through the ribosome, the second group of three N-bases picks up its corresponding tRNA. The amino acid attached to the second tRNA bonds to the one attached to the first tRNA and then the third one bonds to the second, and so on until a chain of amino acids (a protein) is formed, as in the following figure:

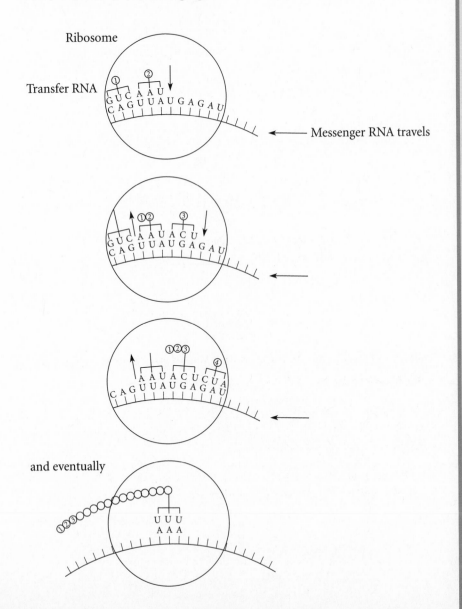

3

463

tRNA binds to mRNA at three consecutive sites, as in the following figure:

When tRNA binds to mRNA, how many bonding sites are there?

_____

These three binding sites are called the *triplet code* or *codon*.

UGG

464

In the diagram in problem 463, the beginning triplet code ACC corresponds to which code in tRNA? _____

mRNA
A C G C U U

tRNA
U G C G A A
Amino acid Amino acid

mRNA
G C A U C C

tRNA
C G U A G G
Amino acid Amino acid

465

Indicate the tRNAs and binding needed for the following segment of mRNA.

THE GENETIC CODE

Each tRNA is specific for a certain amino acid. The following chart indicates the amino acids corresponding to the triplet code in mRNA. Note that there is some redundancy; some amino acids are coded for more than one codon.

Messenger RNA

First Letter	Second Letter	Third Letter			
		A	C	G	U
A	A	lys	asn	lys	asn
	C	thr	thr	thr	thr
	G	arg	ser	arg	ser
	U	ile	ile	met**	ile
C	A	gln	his	gln	his
	C	pro	pro	pro	pro
	G	arg	arg	arg	arg
	U	leu	leu	leu	leu
G	A	glu	asp	glu	asp
	C	ala	ala	ala	ala
	G	gly	gly	gly	gly
	U	val	val	val	val
U	A	*	tyr	*	tyr
	C	ser	ser	ser	ser
	G	*	cys	trp	cys
	U	leu	phe	leu	phe

*Chain terminator
**Chain initiator

466

Look at the answer in problem 465. The first tRNA bonds to the coded group ACG in mRNA. According to the preceding table, ACG in the mRNA corresponds to which amino acid? _____

thr

leu; ala; ser

467

The second tRNA in problem 465 corresponds to which amino acids? _____ The third tRNA? _____ The fourth? _____

The DNA segment above calls for four amino acids that are bonded in this order:

thr-leu-ala-ser

lys-arg-glu-ser

468

Indicate the amino acid chain designated by the following strand of mRNA:

A   A   A   C   G   G   A   G   U   C   C

_____

AUG; UAA, UAG, UGA

469

Which coded group(s) indicate the start of an amino acid chain? _____ The termination of the chain? _____

ACC GCU UUU;
thr-ala-phe

470

Look at the following segment of DNA:

T   G   G   C   G   A   A   A

What is the sequence of mRNA for this segment: _____

What is the sequence of amino acids indicated? _____

MUTATIONS

471

Suppose that somehow one of the N-bases in the DNA in problem 470 was miscopied. Two such examples follow. In the first example, the sixth base is copied as a G instead of a U and in the second example, the last base is copied as an A instead of a U.

mRNA first miscopy: ACC GCG UUU

mRNA second miscopy: ACC GCU UUA

What sequence of amino acids would be indicated by the first miscopy? _____ By the second miscopy? _____

thr-ala-phe; thr-ala-leu

472

A change in copying DNA is called a mutation. Some mutations are not harmful; others might be. Would the change in the first miscopy cause a harmful mutation? _____ In the second miscopy? _____

no; maybe

GAA or GAG

473

The following diagram indicates the sequence of the 146 amino acids present in hemoglobin.

```
1              5                  10                 15
val - his - leu - thr - pro ─┤ glu ├─ glu - lys - ser - ala - val - thr - ala - leu - try - gly - lys ─┐
                                                                                                          \
                  30                 25                 20                                              val
     ┌── val - val - leu - leu - arg - gly - leu - ala - glu - gly - gly - val - glu - asp - val - asn ──┘
     35
    tyr
     └── pro - try - thr - gln - arg - phe - phe - glu - ser - phe - gly - asp - leu - ser - thr - pro ──┐
                  40                 45                 50                                                 \
                  65                 60                 55                                               asp
          ┌────── leu - val - lys - lys - gly - his - ala - lys - val - lys - pro - asn - gly - met - val - ala ─┘
        gly
          └── ala - phe - ser - asp - gly - leu - ala - his - leu - asp - asn - leu - lys - gly - thr - phe ──┐
              70                 75                 80                 85                                      \
                  100                95                 90                                                    ala
          ┌────── asn - glu - pro - asp - val - his - leu - lys - asp - cys - his - leu - glu - ser - leu - thr ─┘
        phe
          └── arg - leu - leu - gly - asn - val - leu - val - cys - val - leu - ala - his - his - phe - gly ──┐
              105                110                115                                                       \
                                                                                                            120
              135                130                125                                                     lys
          ┌── val - gly - ala - val - val - lys - gln - tyr - ala - ala - gln - val - pro - pro - thr - phe - glu ─┘
        ala
          └── asn - ala - leu - ala - his - lys - tyr - his
              140                146
```

Note that the sixth amino acid is glu. What is(are) the coded group(s) for glu? _____

GUA or GUG

474

Suppose that the middle section of either of these code groups was changed from A to U. Then the new coded groups would be either _____ or _____.

<u>475</u>

What amino acid(s) would this new coded group designate?

_____

val

<u>476</u>

Would this new protein of 146 amino acids still be hemoglobin?

_____

no

The new protein, with only one of 146 amino acids changed, is called hemoglobin-S and causes the genetic disease sickle-cell anemia.

Enzymes 16

Enzymes are proteins that act as biological catalysts. A catalyst increases the rate of a reaction but is not changed itself.

Pepsin is an enzyme found in gastric juice. It increases the rate of reaction for the digestion of protein in the stomach. Note that pepsin is an enzyme for a particular substance: protein. In general, enzymes are highly specific; catalysts are nonspecific.

477

Enzymes are _____ (specific/nonspecific) catalysts.

specific

478

a. What effect does an enzyme have on the rate of a reaction?
_____

b. Is an enzyme changed by the reaction? _____

a. it increases it
b. no

NAMES OF ENZYMES

a. carbohydrates
b. lipids

479

Most enzymes have names ending in *ase*. Some enzymes are named for the type of reaction they catalyze. *Hydrolases* are enzymes that catalyze hydrolysis reactions. *Oxidases* are enzymes that catalyze oxidation reactions.

a. Carbohydrases are enzymes that catalyze the hydrolysis of _____.

b. Lipases are enzymes that catalyze the hydrolysis of _____.

a. lactose
b. maltase

480

Other enzymes are named according to the substrate upon which they act. *Sucrase* is an enzyme that catalyzes the hydrolysis of sucrose.

a. Lactase is an enzyme that catalyzes the hydrolysis of _____.

b. Which enzyme catalyzes the hydrolysis of maltose? _____

COENZYMES

Some enzymes are proteins only. Other enzymes have a protein part and a nonprotein part, both of which must be present before the enzyme can function. The protein part of such an enzyme is called the *apoenzyme*; the nonprotein part is called the *coenzyme*.

$$\text{apoenzyme} + \text{coenzyme} \longrightarrow \text{enzyme}$$

Some vitamins act as coenzymes. Nicotinamide adenine dinucleotide, NAD, acts as a coenzyme in oxidation-reduction reactions in the mitochondria. Coenzyme A (CoA) functions in the metabolism of carbohydrates, lipids, and proteins.

481

a.   The nonprotein part of an enzyme is called the _____.

b.   The protein part of an enzyme is called the _____.

482

One example of a coenzyme is _____.

MODE OF ENZYME ACTIVITY

Enzymes are specific catalysts that contain an "active site," that section of the enzyme on which combination with the substrate takes place.

An older theory of enzyme activity, the *lock and key model*, assumes that the active site of an enzyme is rigid and that substrate molecules must fit into that site in order to react. Although it doesn't look like this in reality, one small part of it might be represented like this:

Enzyme

If two compounds, A and B, approach the enzyme, they can attach themselves to it, then react together and go off as a compound, AB. The enzyme is left free to react again and is not used up.

specific; increase

483

Enzymes are _____ (specific/nonspecific) catalysts that _____ the rate of chemical reaction in living systems.

A newer version of the mode of enzyme activity is the *induced fit theory*. In this theory the active site is flexible rather than rigid, and it changes its conformation upon binding to the substrate, thus yielding an enzyme-substrate combination.

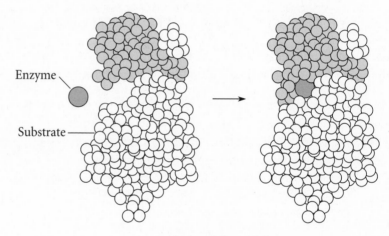

Induced fit model

484

a. rigid
b. flexible

a. In the lock and key model, the active site is _____ (flexible/rigid).

b. In the induced fit model, the active site is _____ (flexible/rigid).

INHIBITORS

485

If some substance other than the substrate fits into part of the active site, can the substrate then react with that enzyme? _____

no

486

A substance that competes for a position in the active site of an enzyme is called a *competitive inhibitor*. Which of the following substances could act as a competitive inhibitor for the active site shown in problem 487?

a   b   c

a

487

If substance **a** reacts with the active site of the enzyme shown here, could that enzyme then function normally? _____

Enzyme

no

**it should inhibit
the growth of those
bacteria**

488

An example of a competitive inhibitor is sulfanilamide, whose structure is similar to that of p-aminobenzoic acid, a compound that is essential for the growth of certain bacteria.

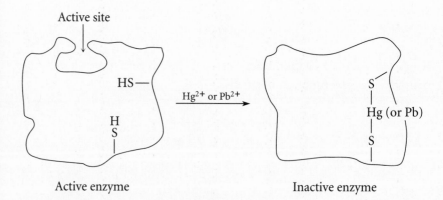

Sulfanilamide   p-Aminobenzoic acid

What effect should sulfanilamide, a competitive inhibitor, have on the growth of the bacteria that need p-aminobenzoic acid? _____

no

489

Consider these diagrams:

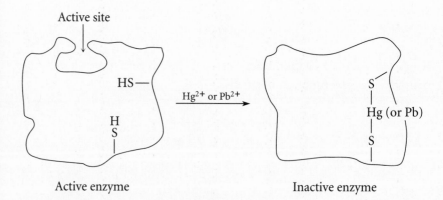

Active enzyme   Inactive enzyme

Note that the shape of the active site has been changed. Can the enzyme then function normally? _____

490

A substance that changes the shape of the active site by reacting with the enzyme at some point other than the active site is called a *noncompetitive inhibitor*. What substance acts as a noncompetitive inhibitor in the reaction in problem 489? _____

Hg^{2+} or Pb^{2+}

491

a. A substance that fits into the active site of an enzyme and prevents that enzyme from functioning normally is called a(n) _____.

b. A substance that changes the shape of the active site of an enzyme by reacting with that enzyme at a position other than the active site is called a(n) _____.

a. competitive inhibitor
b. noncompetitive inhibitor

Biologic Oxidation 17

Oxidation may be defined as:

- 🌿 a loss of an electron (or electrons)

- 🌿 a gain of oxygen

- 🌿 a loss of hydrogen

- 🌿 an increase in oxidation number

The compound lactic acid may be oxidized to pyruvic acid according to the following equation:

Lactic acid   Pyruvic acid

$$\boxed{492}$$

This is an oxidation reaction because it involves:

___ the gain of oxygen by the lactic acid molecule
___ the loss of hydrogen by the lactic acid molecule

the loss of hydrogen by the lactic acid molecule

a. alcohol and acid
b. ketone and acid

493

a.   Lactic acid contains what functional groups? _____

b.   Pyruvic acid contains what functional groups? _____

gain of oxygen

494

Another example of oxidation is the reaction of acetaldehyde to form acetic acid:

$$H-\underset{\underset{H}{|}}{\overset{\overset{H}{|}}{C}}-\overset{\overset{H}{|}}{C}=O \xrightarrow{[O]} H-\underset{\underset{H}{|}}{\overset{\overset{H}{|}}{C}}-\overset{\overset{OH}{|}}{C}=O$$

Acetaldehyde   Acetic acid

This reaction is oxidation because it involves _____.

There is a loss of an electron.

495

The Fe^{2+} ion may be oxidized to the Fe^{3+} ion as follows:

$$Fe^{2+} - e^- \longrightarrow Fe^{3+}$$

Why is this oxidation? _____

Reduction is the opposite of oxidation. Reduction is

✺   a gain of an electron (or electrons)

✺   a gain of hydrogen

✺   a loss of oxygen

✺   a decrease in oxidation number

Oxidation and reduction always occur together. One cannot occur without the other.

496

Consider the following reaction:

COOH   FAD   FADH$_2$ COOH
| |
H—C—H   ⟶   C—H
| ‖
H—C—H   C—H
| |
COOH   COOH

        Succinic acid   Fumaric acid

a.   Is succinic acid oxidized or reduced? _____
     Why? _____

b.   Is FAD oxidized or reduced? _____ Why? _____

Note that both oxidation and reduction are involved in this reaction.

497

Consider the following reaction:

        NADH   NAD$^+$

        FMN   FMNH$_2$

What was oxidized? _____ Why? _____

What was reduced? _____ Why? _____

CELLULAR RESPIRATION

Cellular respiration enables cells to produce energy from food through oxidation-reduction reactions.

Look at Figure A in the Appendix and refer to it for problems 498–521. Note that cellular respiration involves three distinct stages: glycolysis, the Krebs cycle, and the electron transport chain.

cytoplasmic fluid; mitochondrion; mitochondrion

498

In which part of the cell does glycolysis take place? _____

In which part of the cell does the Krebs cycle occur?_____

In which part of the cell does the electron transport chain function? _____

pyruvic acid; cytoplasmic fluid

499

Glycolysis refers to the conversion of glucose to _____ and takes place in which part of the cell? _____

Krebs; mitochondrion

500

Pyruvic acid from the glycolysis sequence is changed to acetyl CoA and enters the _____ cycle, which takes place in the _____ of the cell.

NADH

501

What substance transfers electrons from the glycolysis sequence to the electron transport chain? _____

502

Where does the NADH from the glycolysis sequence go? _____

to the electron
transport chain

503

Where else is NADH produced? _____

in the Krebs cycle and
in conversion of pyruvic
acid to acetyl CoA

504

During glycolysis, glucose is converted to _____.

pyruvic acid

505

Glycolysis takes place in the _____ of the cell.

cytoplasmic fluid

506

In glycolysis, _____ is converted to pyruvic acid.

glucose

507

Electrons are transferred from the glycolysis sequence to the _____
via what substance? _____

electron transport chain;
NADH

508

The electron transport chain functions in which part of the cell?

_____

mitochondrion

mitochondrion

509

The Krebs cycle functions in which part of the cell? _____

3

510

Glycolysis produces pyruvic acid, NADH, and one other substance, ATP, the cell's main energy compound.

The abbreviation ATP refers to adenosine triphosphate. This name indicates that the compound adenosine is bonded to _____ (how many?) phosphate groups.

2

511

The abbreviation ADP refers to adenosine diphosphate, indicating that the adenosine molecule is bonded to _____ (how many?) phosphate groups.

adenosine; 1; phosphate

512

The abbreviation AMP refers to adenosine monophosphate and indicates that the _____ molecule is bonded to _____ (how many?) _____ group(s).

acetyl CoA

513

Pyruvic acid from the glycolysis sequence enters the Krebs cycle after being changed into _____.

514

Electrons are transported from the Krebs cycle via _____ and _____.

NADH; FADH$_2$

515

Electrons are transported from the Krebs cycle to the _____.

electron transport chain

516

The electron transport chain functions in which part of the cell? _____

mitochondrion

517

NADH is produced in three different sequences during cellular respiration. They are _____, _____, and _____.

glycolysis; conversion of pyruvic acid to acetyl CoA; Krebs cycle

518

FADH$_2$ is produced during which sequence? _____

Krebs cycle

519

The Krebs cycle produces the energy compound _____.

ATP

520

The electron transport chain produces the energy compound _____.

ATP

glycolysis; Krebs cycle; electron transport chain

521

During cellular respiration, ATP is produced in the following sequence: _____, _____, and _____.

GLYCOLYSIS

Refer to Figure B in the Appendix for problems 522–541.

glucose; 6

522

Glycolysis begins with _____ (which compound?), which contains _____ (how many?) carbon atoms.

pyruvic acid; 3

523

The end product of glycolysis is _____, which contains _____ (how many?) carbon atoms.

2

524

For each molecule of glucose that goes through the glycolysis sequence, how many molecules of pyruvic acid are produced? _____

cytoplasmic fluid

525

In which part of the cell does glycolysis take place? _____

ADP

526

Note that in the first half of the glycolysis sequence ATP is used up and is converted to _____.

527

How many ATPs are required for the first three steps in the glycolysis sequence? _____

The ATP thus used is converted into _____.

2; ADP

528

In the second half of the glycolysis sequence, energy is produced and ADP is converted to _____.

ATP

529

In the second half of the glycolysis sequence, _____ (how many?) ADPs are converted to ATPs.

4

530

In the first half of the glycolysis sequence, _____ (how many?) ATPs are converted to ADPs.

2

531

In the entire glycolysis sequence, the net number of ATPs produced is _____.

2

2

NAD$^+$

pyruvic acid

2

ATP

532

In addition to ATP, another energy compound, NADH, is produced during glycolysis. How many NADHs are produced from one molecule of glucose during the glycolysis sequence? _____

533

NADH is produced from which compound? _____

534

The overall glycolysis sequence may be written as:

glucose ⟶ _____

535

The overall glycolysis sequence may be further written as:

glucose ⟶ _____ (how many?) pyruvic acid

536

A further addition to this equation is:

ADP   _____

glucose ⟶ 2 pyruvic acid

537

This equation may then be modified to:

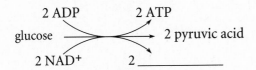

2 ADP _____ (how many?) ATP

glucose ⟶ 2 pyruvic acid

538

This equation may be further modified to:

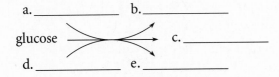

2 ADP   2 ATP

glucose ⟶ 2 pyruvic acid

2 NAD⁺   2 _____

539

The complete glycolysis sequence may thus be written as:

a. _____   b. _____

glucose ⟶ c. _____

d. _____   e. _____

540

A biochemical reaction that requires oxygen is called *aerobic*; one that does not require oxygen is termed *anaerobic*.

Do any of the reactions in the glycolysis sequence require oxygen?

_____

2

NADH

a. 2 ADP
b. 2 ATP
c. 2 pyruvic acid
d. 2 NAD⁺
e. 2 NADH

no

anaerobic

541

Is glycolysis an aerobic or an anaerobic sequence? _____

CONVERSION OF PYRUVIC ACID
TO ACETYL COENZYME A

Refer to Figure C in the Appendix for problems 542–555.

pyruvic acid

542

The end product of the glycolysis sequence is _____.

cytoplasmic fluid

543

Pyruvic acid is produced in which part of the cell? _____

mitochondrion

544

Pyruvic acid from the glycolysis sequence diffuses from the cytoplasmic fluid into which part of the cell? _____

no

545

Does pyruvic acid enter the Krebs cycle directly? _____

546

Pyruvic acid is converted into which compound in the mitochondrion?

_____

acetyl CoA

547

Does acetyl CoA (acetyl coenzyme A) enter the Krebs cycle? _____

yes

548

Each molecule of glucose that enters the glycolysis sequence produces
_____ (how many?) molecule(s) of pyruvic acid.

2

549

Each molecule of pyruvic acid that diffuses from the cytoplasmic
fluid into the mitochondrion produces _____ (how many?)
molecule(s) of acetyl CoA.

1

550

Therefore, during glycolysis, one molecule of glucose produces
_____ molecule(s) of pyruvic acid, which in turn are used
to produce _____ molecule(s) of acetyl CoA.

2; 2

6; 6; 2

551

One molecule of glucose contains _____ (how many?) carbon atoms.

Two molecules of pyruvic acid contain _____ (how many?) carbon atoms.

The acetyl groups of two molecules of acetyl CoA contain four carbon atoms.

How many carbon atoms are missing during this overall set of reactions? _____

2; pyruvic acid

552

As is shown in Figure C, in the cytoplasmic fluid, one molecule of carbon dioxide is produced for each molecule of pyruvic acid converted into acetyl CoA. Thus, one molecule of glucose enters the glycolysis sequence and is converted to _____ (how many?) molecule(s) of _____.

2; acetyl CoA

553

In the mitochondrion, two molecules of pyruvic acid are converted into _____ (how many?) molecule(s) of _____, which then enter the Krebs cycle.

carbon dioxide or CO_2

554

The missing carbon atoms show up as _____.

555

For each molecule of acetyl CoA produced from pyruvic acid, _____ (how many?) molecule(s) of NADH are produced. How many molecules of CO_2 are produced? _____

1; 1

THE KREBS CYCLE

Refer to Figure D in the Appendix for problems 556–566.

556

The two carbon atoms of the acetyl part of acetyl CoA participate in the Krebs cycle while the CoA molecule is recycled.

The two carbon atoms from the acetyl group are eliminated from the Krebs cycle as two molecules of what compound? _____

CO_2

557

As one molecule of acetyl CoA goes through the Krebs cycle:

How many molecules of ATP are produced? _____

How many molecules of NADH are produced? _____

How many molecules of $FADH_2$ are produced? _____

1; 3; 1

558

For every molecule of acetyl CoA that enters the Krebs cycle, _____ molecule(s) of ATP, _____ molecule(s) of NADH, and _____ molecule(s) of $FADH_2$ are produced.

1; 3; 1

ADP

NAD$^+$

FAD

mitochondrion

CO$_2$; ATP; NADH; FADH$_2$

no

anaerobic

559
In the Krebs cycle, ATP is produced from what compound? _____

560
In the Krebs cycle, NADH is produced from what compound? _____

561
In the Krebs cycle, FADH$_2$ is produced from what compound? _____

562
In which part of the cell does the Krebs cycle take place? _____

563
The four products eliminated from the Krebs cycle are: _____, _____, _____, and _____.

564
Is oxygen required in any of the steps of the Krebs cycle? _____

565
Is the Krebs cycle an aerobic or anaerobic sequence? _____

566

Where do the NADH and the $FADH_2$ produced by the Krebs cycle go?

_____

to the electron
transport chain

ELECTRON TRANSPORT CHAIN

Refer to Figure E in the Appendix for problems 567–578.

567

The reactions in the electron transport chain take place in which part of
the cell? _____

mitochondrion

568

Is oxygen required in any of the steps of the electron transport chain?

_____

yes

569

Is the electron transport chain an aerobic or anerobic sequence?

_____

aerobic

570

Which compounds enter the electron transport chain? _____

NADH, $FADH_2$

571

The $FADH_2$ entering the electron transport chain comes from which
sequence? _____

Krebs cycle

glycolysis, conversion of pyruvic acid to acetyl CoA, and Krebs cycle

572

The NADH entering the electron transport chain comes from three different sequences. They are _____
_____.

ATP

573

What is the primary energy compound produced by the electron transport chain? _____

yes

574

Is ATP produced during glycolysis? _____

no

575

Is ATP produced during the conversion of pyruvic acid to acetyl CoA?
_____

yes

576

Is ATP produced in the Krebs cycle? _____

yes

577

Is ATP produced in the electron transport chain? _____

ADP

578

ATP is produced from which compound? _____

Photosynthesis 18

OVERALL REACTION

| 579 |

The overall reaction of photosynthesis is:

$$6CO_2 + 6H_2O \xrightarrow[\text{chlorophyll}]{\text{light}} C_6H_{12}O_6 + 6O_2$$

During photosynthesis,

a. the raw materials are _____

b. the end products are _____

c. the source of energy is _____

d. the source of CO_2 is _____

e. the source of H_2O is _____

f. what happens to the O_2? _____

a. CO_2 and H_2O
b. $C_6H_{12}O_6$ and O_2
c. light
d. air
e. the ground
f. released into air

animal; plant

580

Look at the following structures for heme (which is a part of the hemoglobin molecule) and chlorophyll *a*:

Heme

Chlorophyll *a*

Note that these two substances have similar structures. However, heme (hemoglobin) is necessary for most _____ (animal/plant) life while chlorophyll *a* is necessary for most _____ (animal/plant) life.

581

What metal is at the center of the heme molecule? The chlorophyll *a* molecule? _____

Fe; Mg

582

Consider the following two examples of the overall photosynthesis reaction. Radioactive oxygen, ^{18}O, is introduced in the form of radioactive $C^{18}O_2$ in the first reaction and as radioactive water, $H_2^{18}O$, in the second reaction.

$$6C^{18}O_2 + 6H_2O \xrightarrow[\text{chlorophyll}]{\text{light}} C_6H_{12}^{18}O_6 + 6O_2$$

$$6CO_2 + 6H_2^{18}O \xrightarrow[\text{chlorophyll}]{\text{light}} C_6H_{12}O_6 + 6^{18}O_2$$

What is the source of the O_2 produced during photosynthesis, CO_2 or H_2O?

H_2O

583

Photosynthesis may be subdivided into two main steps as indicated in the following simplified chart.

Photosynthesis

a. light
b. H_2O
c. ATP, NADPH; O_2
d. H_2O

For the first step, *light reactions*,

a. what is the energy source? _____

b. what reactant is necessary? _____

c. what compounds are produced? _____

d. where does the O_2 come from? _____

584

a. CO_2
b. carbohydrate or
   $C_6H_{12}O_6$
c. ADP; $NADP^+$

For the second step, the Calvin cycle,

a. what is the principal reactant? _____

b. what is the principal product? _____

c. what energy compounds are returned to the light reactions step?

_____

585

no

Is light required for the Calvin cycle? _____

LIGHT REACTIONS

The following is a simplifed diagram of light reactions.

586

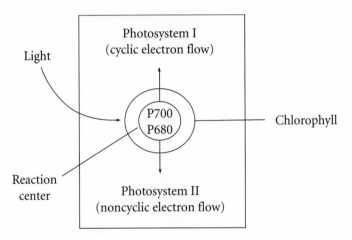

When light strikes the chlorophyll molecules, it excites an electron, which is transferred from chlorophyll molecule to chlorophyll molecule until it reaches the reaction center. In photosystem I, the reaction center is associated with a specific protein (pigment) known as P700 because it is most effective at absorbing light with a wavelength of 700 nm. Photosystem II is associated with a similar protein P680 that is sensitive to absorbing light with a _____.

wavelength of 680 nm

587

Which photosystem has a noncyclic electron flow? _____

II

P700; P680

What numbered form of chlorophyll *a* is associated with photosystem I?
_____ With photosystem II? _____

I; II

Which photosystem is cyclic? _____ Noncyclic? _____

The following is an abbreviated diagram of photosystem I.

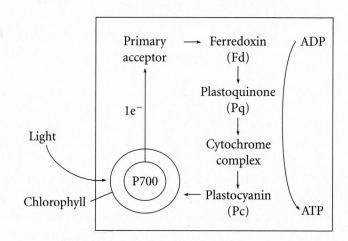

1

How many electrons are involved in this system? _____

yes

In photosystem I, an electron is emitted by P700 and travels to a primary
acceptor. After a series of reactions, does the electron return to P700?

_____

This is why this system is called cyclic.

592

Is H_2O split to provide electrons in photosystem I? _____

no

593

Is O_2 produced during the reactions of photosystem I? _____

no

594

What product is the result of the reactions in photosystem I? _____

ATP

595

In photosystem I, ATP is produced from _____.

ADP

596

When ADP is converted to ATP, what group must be added? _____

P (or phosphate)

597

The production of ATP during the cyclic electron flow is called *cyclic photophosphorylation*.

During cyclic photophosphorylation, _____ is converted to _____.

ADP; ATP

no

598

Is NADPH produced in this sequence? _____

no

599

Is O_2 produced in this sequence? _____

P700; light

600

The energy for the electron involved in photosynthesis I is produced from the chlorophyll *a* that is labeled _____ and in turn is produced by the action of _____ on chlorophyll.

The following is an abbreviated diagram of photosystem II.

yes; no;
yes; yes

601

In photosystem I, is ATP produced? _____ NADPH? _____
_____

In photosystem II, is ATP produced? _____ NADPH? _____
_____

602

Where do the electrons used by photosystem II come from?

_____

water

603

When the H_2O gives off electrons, what substances are produced?

_____

H^+, O_2

604

Is the same type of chlorophyll _a_ needed for both photosystems?

_____

no, system I needs P700 and system II needs P680

605

Can ATP be produced in both photosystems? _____

yes

606

Is photosystem II cyclic? _____

no

607

Is water splitting required for photosystem I? _____
Photosystem II? _____

no; yes

608

Is O_2 produced in photosystem I? _____ In photosystem II?

_____

no; yes

no; yes

$$609$$

Is NADPH produced in photosystem I? _____ In photosystem II? _____

NADP$^+$

$$610$$

From what substance is NADPH produced? _____

H$_2$O

$$611$$

From what substance is O_2 produced? _____

The light reactions system may be abbreviated as follows:

light

$$612$$

What is the original energy source for this series of reactions?

_____

613

When light strikes photosystem I and photosystem II, it causes electrons to flow.

Which of these systems is cyclic? _____

Which is noncyclic? _____

I; II

614

In which photosystem is water splitting involved? _____

II

615

What is the source of oxygen during the light reactions? _____

water

616

By which photosystem is oxygen produced? _____

II

617

What high-energy compound(s) is(are) produced in photosystem I? _____

ATP

618

What high-energy compound(s) is(are) produced in photosystem II? _____

ATP, NADPH

ADP

NADP$^+$

3

619

From what substances is ATP produced? _____

620

From what substance(s) is NADPH produced? _____

CALVIN CYCLE

The following is an abbreviated diagram of the Calvin cycle.

621

The end product of the Calvin cycle is glyceraldehyde-3-P. It contains how many carbon atoms? _____

622

CO_2 contains how many carbon atoms? _____

1

623

How many CO_2 molecules must go through the Calvin cycle to produce one molecule of glyceraldehyde-3-P? _____

3

624

Is CO_2 a part of the light reactions sequence or the Calvin cycle? _____

Calvin cycle

625

What high-energy compounds enter the Calvin cycle? _____

ATP, NADPH

626

Where do these energy compounds come from? _____

light reactions

The following is a simplified illustration of the Calvin cycle. Numbers in brackets [ ] indicate total number of carbon atoms. Enzymes for each step are omitted for simplicity.

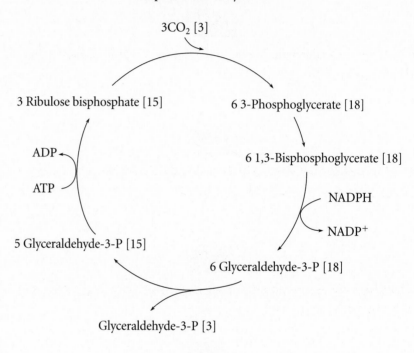

Simplified Calvin cycle

3

627

How many carbon atoms enter the cycle? _____

CO_2

628

In what compound do they enter? _____

glyceraldehyde-3-P

629

In what compound do the carbons exit the Calvin cycle? _____

630

Glyceraldehyde-3-P contains how many carbons? _____

3

631

Does all the glyceraldehyde-3-P produced in the Calvin cycle leave the cycle? _____

no

632

What fraction of the glyceraldehyde-3-P leaves that cycle? _____

⅙

633

Does the number of carbon atoms entering, reacting in, and leaving the cycle balance? _____

yes

634

What energy compounds enter the Calvin cycle? _____

ATP, NADPH

635

What energy compounds are produced in the Calvin cycle? _____

ADP, $NADP^+$

ATP; NADPH

636

These energy compounds return to the light reactions system. There ADP is regenerated to _____ and $NADP^+$ is regenerated to _____.

3

637

Glyceraldehyde-3-P contains how many carbon atoms? _____

6

638

Glucose, $C_6H_{12}O_6$, contains how many carbon atoms? _____

2

639

How many glyceraldehyde molecules must be combined to produce one glucose (carbohydrate) molecule? _____

The following figure gives an overall summary of photosynthesis.

Summary of photosynthesis

640

What is the substance **A** that enters the light reactions system?

_____

641

What is substance **B**? _____

642

What compounds are represented by **E**? _____

643

What compounds are represented by **F**? _____

644

What is compound **C**? _____

645

What is compound **D**? _____

H_2O

O_2

ATP, NADPH

ADP, $NADP^+$

CO_2

glyceraldehyde-3-P

C$_4$ PLANTS

C$_4$ plants are those that have an alternate method of fixing CO_2 for the Calvin cycle as indicated in the following abbreviated diagram. (Numbers in brackets [ ] refer to the number of carbon atoms in each molecule.)

air

646

What is the source of CO_2? _____

malate, oxaloacetate

647

What four-carbon compounds are involved in the C$_4$ cycle?

_____

648

What high-energy compound is needed for the C_4 cycle? _____

ATP

649

When CO_2 enters the Calvin cycle, what high-energy compounds are required? _____

ATP, NADPH

650

Where are these high-energy compounds produced? _____

in light reactions sequence

Oxygen–Carbon Dioxide Transport in Blood 19

FLOW OF GASES

Gases diffuse from an area of higher pressure to one of lower pressure.

In the following diagram P_{O_2} indicates the partial pressure of oxygen (that part of the total pressure exerted by the oxygen) in millimeters of mercury.

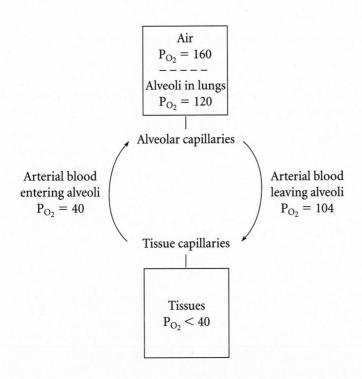

Air
$P_{O_2} = 160$
$- - - - -$
Alveoli in lungs
$P_{O_2} = 120$

Alveolar capillaries

Arterial blood entering alveoli
$P_{O_2} = 40$

Arterial blood leaving alveoli
$P_{O_2} = 104$

Tissue capillaries

Tissues
$P_{O_2} < 40$

into the capillaries in the alveoli

651

Which way does oxygen diffuse in the lungs? _____

gases diffuse from higher to lower partial pressure

652

From the alveoli, oxygen diffuses into the blood and then to the tissues. Why does it move in that direction? _____

picks up

653

When the arterial blood flows back to the alveoli, does it pick up or give off oxygen? _____

Tissues require oxygen for metabolic reactions such as:

$$C_6H_{12}O_6 + 6O_2 \longrightarrow 6CO_2 + 6H_2O$$

Carbon dioxide is a gaseous waste product of oxidative metabolism. Its partial pressures in various parts of the blood system are shown in the following figure.

Air
$P_{O_2} = 160$
$P_{CO_2} = 0.2$
- - - - -
Alveoli in lungs
$P_{O_2} = 120$
$P_{CO_2} = 27$

Arterial blood entering alveoli
$P_{O_2} = 40$
$P_{CO_2} = 45$

Venous blood
$P_{O_2} = 40$
$P_{CO_2} = 45$

Arterial blood leaving alveoli
$P_{O_2} = 104$
$P_{CO_2} = 40$

Tissues
$P_{O_2} < 40$
$P_{CO_2} > 45$

654

From the tissues, which way does carbon dioxide flow? _____

655

Why does carbon dioxide flow in that direction? _____

TRANSPORT OF OXYGEN

656

When oxygen enters the red blood cells, it binds with hemoglobin, here represented by the formula HHb, to form a compound known as oxy-hemoglobin.

$$HHb + O_2 \longrightarrow HHbO_2$$

Hemoglobin   Oxyhemoglobin

Oxyhemoglobin in turn is ionized into the oxyhemoglobin ion and a hydrogen ion.

$$HHbO_2 \longrightarrow H^+ + HbO_2^-$$

Oxygen is carried to the tissues in the form of the oxyhemoglobin ion, or _____.

657

In the tissues, the oxyhemoglobin ion binds with a hydrogen ion to form oxyhemoglobin; the formula is:

$$HbO_2^- + H^+ \longrightarrow \underline{\hspace{3cm}}$$

$HHbO_2 \longrightarrow HHb + O_2$

658

The oxyhemoglobin releases its oxygen to form hemoglobin. This reaction may be written as:

_____

The oxygen thus liberated is used by the tissues for various metabolic reactions. The hemoglobin once again is available to pick up oxygen in the alveoli.

hemoglobin

659

What substance in the blood carries oxygen to the tissues? _____

HHb

660

What is the symbol for hemoglobin? _____

oxyhemoglobin

661

When hemoglobin picks up oxygen, _____ is formed.

$HHbO_2$

662

What is the formula of oxyhemoglobin? _____

$HHb + O_2 \longrightarrow HHbO_2$

663

Write the reaction for hemoglobin and oxygen.

_____

664

Write the formula for the ionization of oxyhemoglobin.

_____

665

In what form is oxygen carried to the tissues? _____

666

Write the reaction of the oxyhemoglobin ion with a hydrogen ion in the tissues.

_____

667

Write the reaction indicating how oxyhemoglobin gives off its oxygen in the tissues.

_____

668

What happens to the hemoglobin after it gives off its oxygen?

_____

$HHbO_2 \rightarrow H^+ + HbO_2^-$

HbO_2^- or oxyhemoglobin ion

$HbO_2^- + H^+ \rightarrow HHbO_2$

$HHbO_2 \rightarrow HHb + O_2$

returns to alveoli to pick up more O_2

red blood cells

669

What part of the blood contains hemoglobin? _____

The overall steps in this process may be diagrammed as follows:

Red blood cells
in lung capillaries   Red blood cells
in tissues

Arterial flow

$HHbO_2 \rightarrow HbO_2^-$ $HbO_2^- \rightarrow HHbO_2$

H^+ H^+

O_2
from
alveoli   HHb   HHb O_2
to
tissues

Venous flow

TRANSPORT OF CARBON DIOXIDE

670

**gases diffuse from an
area of higher to lower
partial pressure**

Carbon dioxide is produced in the tissues and diffuses into the red blood cells.

Why does it diffuse in that direction? _____

671

In the red blood cells, carbon dioxide reacts with water to produce carbonic acid:

$$CO_2 + H_2O \longrightarrow H_2CO_3$$
Carbonic acid

Carbonic acid in turn ionizes, producing a hydrogen ion and a bicarbonate ion.

$$H_2CO_3 \longrightarrow H^+ + HCO_3^-$$
Bicarbonate
ion

Red blood cells are able to hold only a small amount of the bicarbonate ions produced from carbon dioxide, so the excess diffuses into the blood plasma.

The bicarbonate ion has what charge? _____

672

To replace the bicarbonate ions, chloride ions (Cl^-) flow from the plasma into the red blood cells.

Chloride ions have what charge? _____

673

The process of shifting bicarbonate and chloride ions in opposite directions across a red blood cell membrane is called the *chloride shift*. It takes place in the red blood cells in the lung and tissue capillaries.

In the chloride shift, _____ ions flow from the red blood cells to the _____.

-1

-1

HCO_3^-; plasma

674

As bicarbonate ions flow out of the red blood cells, _____ ions flow from the _____ into the _____.

675

In the lungs, the bicarbonate ion combines with a hydrogen ion to re-form carbonic acid.

$$HCO_3^- + H^+ \longrightarrow H_2CO_3$$

Carbonic acid in turn decomposes into water and carbon dioxide, which is then exhaled.

$$H_2CO_3 \longrightarrow H_2O + CO_2$$

676

What happens to the carbon dioxide in the lungs? _____

677

As the bicarbonate ion is used up in the preceding reactions, it is replaced by bicarbonate ions from the plasma. At the same time, chloride ions from the red blood cells flow back into the plasma. This is termed a *reverse chloride shift*.

In a reverse chloride shift, _____ ions flow from the red blood cells into the plasma and at the same time _____ ions flow from the plasma into the red blood cells.

678

These reactions may be diagrammed as follows:

Red blood cells
in lung capillaries

Red blood cells
in tissues

What does part **A** represent? _____

What does part **B** represent? _____

679

In addition to the red blood cells transporting carbon dioxide as bicarbonate ions, some of the carbon dioxide reacts with hemoglobin to form carbaminohemoglobin:

$$CO_2 + HHb \longrightarrow HHbCO_2$$

In the lungs, carbaminohemoglobin decomposes into hemoglobin and carbon dioxide.

What happens to the carbon dioxide? _____

it is exhaled

680

What happens to the hemoglobin? _____

it is recycled

hemoglobin

$$\boxed{681}$$

Oxygen is carried to the tissues by what compound? _____

bicarbonate; carbamino-hemoglobin

$$\boxed{682}$$

Carbon dioxide is carried from the tissues in the form of _____ ions and also in the compound _____.

TRANSPORT OF BOTH OXYGEN AND CARBON DIOXIDE

The interrelationship of the oxygen and carbon dioxide transport systems is shown in the following diagram. Note how the hydrogen ions in the red blood cells in both the lungs and the tissues are recycled.

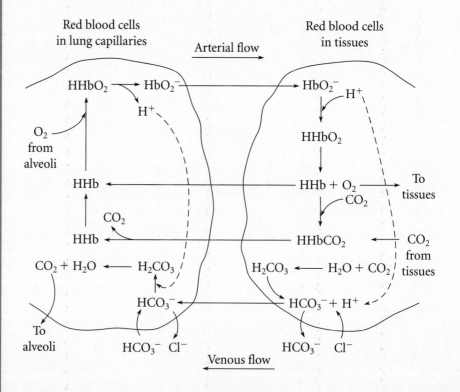

Review

683

What elements do the following symbols represent?

a. H _____   b. O _____   c. S _____

d. Na _____   e. K _____   f. Fe _____

684

Write the symbol for each of the following elements:

a. chlorine _____   b. zinc _____   c. calcium _____

d. carbon _____   e. bromine _____   f. magnesium _____

685

When an atom gains an electron, it forms an ion with what charge?

_____

686

The chlorine atom gains 1 electron to form a chloride ion. The symbol for the chloride ion is _____.

ionic

687

When two ions are held together by the attraction of their opposite charges, what type of bond is between them? _____

hydrogen (or H$^+$)

688

An acid is a substance that yields _____ ions in solution.

hydrogen (or H$^+$)

689

A base is a substance that accepts _____ ions in solution.

any ions except H$^+$
and OH$^-$

690

Salts yield what types of ions? _____

an electrolyte

691

A solution that conducts electricity is called _____.

a nonelectrolyte

692

A solution that does not conduct electricity is called _____.

693

Which of the following are anions? _____

Cations? _____

$$Na^+ \quad Cl^- \quad Mg^{2+}$$
$$SO_4^{2-} \quad K^+ \quad NO_3^-$$

anions: Cl^-, SO_4^{2-}, NO_3^-; cations: Na^+, Mg^{2+}, K^+

694

Saliva has a pH between 5.5 and 6.9. What type of liquid is saliva?

_____

a weak acid

695

Blood has a pH range of 7.35–7.45. What type of liquid is blood?

_____

a weak base

696

An acid of pH 2.75 is how many times as strong as one of pH 3.75?

_____

10

697

A solution that maintains a constant pH upon the addition of either acid or base is called a _____.

buffer

sugar; water

698

In a solution of sugar in water, the solute is _____ and the solvent is _____.

yes; no

699

Do solutions pass through membranes? _____ Do colloids? _____

100

700

A 10% glucose solution will contain 10 g of glucose in _____ mL of solution.

701

1 liter

A 1M KCl solution will contain 1 mole of KCl in how much solution? _____

702

0.5M NaCl

Which solution will have a higher osmolarity, 0.5M NaCl or 0.5M glucose? _____

703

covalent

When two atoms share electrons, they are held together by a _____ bond.

704

solute

Diffusion is the flow of _____ (solute/solvent).

705

Osmosis is the flow of _____ (solute/solvent).

solvent

706

Is osmosis active or passive transport? _____

passive

707

Is diffusion a form of active transport? _____

no

708

A hypertonic solution has a concentration _____ (higher than/lower than/the same as) that of another solution.

higher than

709

An isotonic solution has a concentration _____ (higher than/lower than/the same as) that of another solution.

the same as

710

A hypotonic solution has a concentration _____ (higher than/lower than/the same as) that of another solution.

lower than

711

A selective membrane is _____ (permeable/semipermeable/nonpermeable).

semipermeable

opposite directions

K^+; Na^+

ATP

Na^+ out; K^+ in

3 Na^+ out; 2 K^+ in

−

712

Do osmosis and diffusion normally take place in the same direction or in opposite directions? _____

713

Which ion, Na^+ or K^+, has a higher intracellular concentration in animal cells? _____ Extracellular concentration? _____

714

What high-energy compound frequently supplies the energy for active transport? _____

715

During active transport by the Na/K pump, which ions are transported out of the cell? _____ Into the cell? _____

716

What are the relative numbers of Na^+ and K^+ ions transported out of and into a cell? _____

717

Because of active transport, the interior of the cell has a relative _____ (+/−) charge.

718

When a nerve cell is resting, are its sodium channels normally open or closed? _____ Its potassium channels? _____

both closed

719

When a stimulus affects a resting nerve cell, which channels open?

_____

sodium

720

Which ions move, and do they enter or leave the cell? _____

Na^+ enter

721

When a typical resting nerve cell has a stimulus applied, it becomes _____ (polarized/depolarized).

depolarized

722

After a nerve cell has been depolarized, the _____ channels close and the _____ channels open.

Na^+; K^+

723

This latter change is called _____.

repolarization

K^+

active transport

Ca^{2+}

causes vesicles to fuse to
presynaptic membrane

neurotransmitters;
postsynaptic

They cause ion channels
to open, which in turn
brings about the depolar-
ization of the postsynap-
tic membrane and thus
the transmission of the
impulse along the next
axon.

724

During the undershoot, which channels are slow to close?

_____

725

After the undershoot, which process returns the cell to its resting state?

_____

726

When a new impulse reaches the presynaptic membrane of an axon, it
depolarizes that membrane, causing an influx of _____ ions.

727

The calcium ions have what effect on the vesicles near the presynaptic
membrane? _____

728

When the vesicles fuse with the presynaptic membrane, they release
_____, which travel across the synaptic cleft to the
_____ membrane.

729

What effect do the neurotransmitters have on the postsynaptic
membrane? _____

_____

_____

730

Diagram the structure of the hydrocarbon compound containing the following arrangement of carbon atoms, and indicate all the hydrogen atoms.

$$
\begin{array}{c}
\text{C} \\
| \\
\text{C} - \text{C} - \text{C} \\
| \\
\text{C}
\end{array}
$$

731

Indicate the hydrogen atoms attached to the following arrangement of carbon atoms.

$$ \text{C} \equiv \text{C} - \text{C} $$

732

What is the simplified structure of this compound?

$$
\begin{array}{c}
\text{H} \quad \text{H} \quad \text{H} \quad \text{H} \quad \text{H} \\
| \quad | \quad | \quad | \quad | \\
\text{H} - \text{C} - \text{C} - \text{C} - \text{C} - \text{C} - \text{H} \\
| \quad | \quad | \quad | \\
\text{H} \quad | \quad \text{H} \quad \text{H} \quad \text{H} \\
\text{H} - \text{C} - \text{H} \\
| \\
\text{H}
\end{array}
$$

Answer box:

$$
\begin{array}{c}
\text{H} \\
| \\
\text{H} - \text{C} - \text{H} \\
\text{H} \quad | \quad \text{H} \\
| \quad | \quad | \\
\text{H} - \text{C} - \text{C} - \text{C} - \text{H} \\
| \quad | \quad | \\
\text{H} \quad | \quad \text{H} \\
\text{H} - \text{C} - \text{H} \\
| \\
\text{H}
\end{array}
$$

$$
\begin{array}{c}
\text{H} \\
| \\
\text{H} - \text{C} \equiv \text{C} - \text{C} - \text{H} \\
| \\
\text{H}
\end{array}
$$

$CH_3CH(CH_3)CH_2CH_2CH_3$

733

What is the simplified structure of this compound?

734

Is the water molecule polar or nonpolar? _____

735

What types of bonds are present *between* water molecules? _____

736

What type of bond is involved in maintaining the coil shape of a protein? _____

polar

hydrogen

hydrogen

737

When DNA replicates, it "unzips," breaking what type of bonds?
_____

hydrogen

738

Are hydrogen bonds strong or weak? _____

weak

739

In a protein, hydrogen bonds occur between an H attached to a(n)
_____ atom and an O that is part of a(n) _____
group.

N; C=O

740

What type of functional group is represented by the following?

a. OH _____   b. COOH _____

c. C—C—C _____   d. OPO_3H_2 _____
        ||
        O

e. NH_2 _____   f. COOC _____

g. CHO _____

a. alcohol
b. acid
c. ketone
d. phosphate
e. amine
f. ester
g. aldehyde

Identify the types of functional groups present in each of the compounds in problems 741–743.

acid, amine, disulfide

741

COOH   COOH
| |
NH₂CH   NH₂CH
| |
H₂C—S—S—CH₂

thiol or sulfhydryl

742

$CH_3CH_2CH_2SH$

a. amine and ketone
b. aldehyde and alcohols
c. alcohol and acid

743

a.

NH₂
(benzene ring)
=O

b.

H
|
C=O
|
H—C—OH
|
H—C—OH
|
H—C—OH
|
H—C—OH
|
H

c. $CH_3CH(OH)COOH$

a. _____   b. _____   c. _____

744

Which of the following compounds is a carbohydrate?

__ C_6H_6 __ $C_{12}H_{22}O_{11}$
__ C_6H_7ON __ C_2H_6O

$C_{12}H_{22}O_{11}$

745

a. Monosaccharides have names ending in _____.

b. Most enzymes have names ending in _____.

a. *ose*
b. *ase*

746

An example of a pentose is _____; of a hexose is _____.

ribose; glucose

747

The prefix *deoxy* means _____.

with one less oxygen

748

Hydrolysis of a disaccharide yields _____.

two monosaccharides

glycosidic

polysaccharide

a. monosaccharides
b. polysaccharides

oils

fatty acids; glycerol

749

The bond holding the two halves of a disaccharide together is
called a _____ bond.

750

Starch is a _____ (monosaccharide/disaccharide/
polysaccharide).

751

a.   Which type of carbohydrate can be absorbed directly into the
     blood? _____

b.   Which type of carbohydrate forms a colloidal dispersion in
     water? _____

752

Which are unsaturated, fats or oils? _____

753

The products of fat hydrolysis are _____ and _____.

754

A phospholipid, on hydrolysis, yields _____, _____, _____, and _____.

fatty acids; glycerol; phosphoric acid; a nitrogen compound

755

An example of a steroid is _____.

cholesterol or bile salts or sex hormones

756

Proteins always contain which elements? _____

C, H, O, N

757

The hydrolysis of protein yields _____.

amino acids

758

A polypeptide contains _____ amino acids held together by _____ bonds.

many; peptide

a; b and c; d; e

759

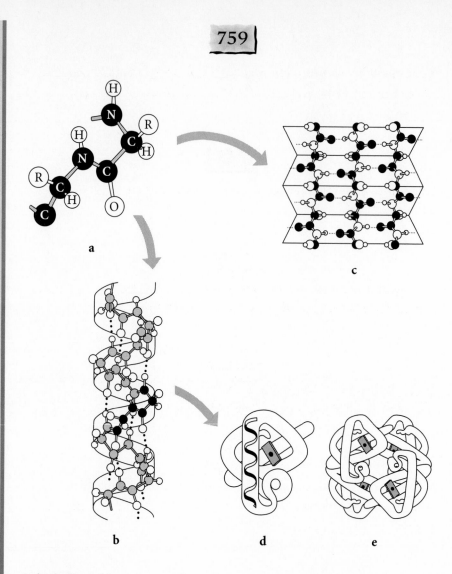

a

b

c

d

e

Which figure(s) represents the primary structure of a protein?

_____

Which figure(s) represents the secondary structure of a protein?

_____

Which figure(s) represents the tertiary structure of a protein?

_____

Which figure(s) represents the quaternary structure of a protein?

_____

760

A tripeptide consists of _____ (how many?) amino acids linked together by what type of bond? _____

3; amide or peptide

761

Nucleotides may contain one of two pentoses. They are _____ and _____.

ribose; deoxyribose

762

The compound that provides chemical energy for the body is _____.

ATP (or adenosine triphosphate)

763

ATP contains how many high-energy phosphate bonds? _____

2

764

The opposite sides of the DNA coils are held together by _____ bonds.

hydrogen

a. T
b. U
c. G
d. G

C   A   G   T   T   C
||| || ||| || || |||
G   U   C   A   A   G

val, lys

a. In DNA, the partner of A is _____.

b. As RNA is synthesized, the presence of A in DNA will specify the incorporation of _____ in the RNA.

c. As RNA is synthesized, the presence of C in DNA will specify the incorporation of _____ in the RNA.

d. In DNA, the partner of C is _____.

766

Complete the following diagram indicating partners and bonds in the formation of mRNA.

DNA

C   A   G   T   T   C

mRNA   _____

767

Which amino acids are designated by the mRNA in problem 766? (See table on page 201.) _____

768

If a genetic code is miscopied, such a change is called a(n) _____.

769

Are all mutations harmful? _____

770

Enzymes are _____ (specific/nonspecific).

771

The nonprotein part of an enzyme is called the _____.

772

a. A substance that "fits" into the active site of an enzyme and prevents that enzyme from functioning is called a _____.

b. A substance that alters the shape of the "active" site of an enzyme by reacting with that enzyme at a spot other than the active site is called a _____.

mutation

no

specific

coenzyme

a. competitive inhibitor
b. noncompetitive inhibitor

succinic acid; FAD

773

Which of the two acids in the diagram is oxidized? _____

Which substance is reduced? _____

COOH
|
CH_2
|
CH_2
|
COOH

Succinic acid

FAD

COOH
|
CH
||
CH
|
COOH

Fumaric acid

$FADH_2$

pyruvic acid

774

Glycolysis involves the conversion of glucose to _____ .

cytoplasmic fluid

775

Glycolysis takes place in which part of the cell? _____

anaerobic

776

Is glycolysis an aerobic or anaerobic sequence? _____

777

Does glycolysis produce ATP? _____

778

The NADH produced during glycolysis is transferred to which sequence? _____

779

Pyruvic acid is converted into _____ in which part of the cell? _____

780

Acetyl CoA enters the _____ cycle, which takes place in which part of the cell? _____

781

Is the Krebs cycle an aerobic or anaerobic sequence? _____

782

Is ATP produced during the Krebs cycle? _____

electron transport chain

783

The NADH and $FADH_2$ produced during the Krebs cycle are transferred to which sequence? _____

mitochondrion

784

The electron transport chain occurs in which part of the cell? _____

aerobic

785

Is the electron transport chain an aerobic or anaerobic sequence? _____

glycolysis, Krebs cycle, electron transport chain

786

In which parts of the cellular respiration sequence is ATP produced?

_____

light

787

What is the source of energy for photosynthesis? _____

carbon dioxide, water

788

Which reactants are involved in photosynthesis? _____

water

789

From which compound is oxygen produced? _____

790

Which photosystem can be cyclic? _____ Which is noncyclic?
_____

I; II

791

Which compounds are produced in the light reactions? _____

ATP, NADPH, O_2

792

ATP is produced from which compound? _____

ADP

793

NADP is produced from which compound? _____

$NADP^+$

794

Which compound enters the Calvin cycle? _____

carbon dioxide

795

Which compound is produced by the Calvin cycle? _____

glyceraldehyde-3-P

796

Which high-energy compounds are produced by the Calvin cycle?
_____

ADP, $NADP^+$

no; yes

797

Is oxygen produced in light reactions photosystem I? _____ In light reactions photosystem II? _____

higher; lower

798

Gases flow by diffusion from an area of _____ (higher/lower) partial pressure to an area of _____ (higher/lower) partial pressure.

oxygen; carbon dioxide

799

In the lungs arterial blood picks up _____ gas and gives off _____ gas.

hemoglobin

800

What substance in a red blood cell transports oxygen? _____

oxyhemoglobin

801

When hemoglobin reacts with oxygen, what compound is formed? _____

yes

802

Is hemoglobin recycled? _____

803

Carbon dioxide is a waste product produced in the tissues. It flows into the red blood cells, where it reacts with water to form _____.

804

Carbonic acid ionizes to form the _____ ion and the _____ ion.

805

The flow of bicarbonate ions from the red blood cells into the blood plasma and the opposite flow of chloride ions from the plasma into the red blood cells are called the _____.

806

Where in the body does the reverse chloride shift take place?

_____

807

Carbon dioxide is carried in the blood in two forms, the _____ ion and the compound _____.

carbonic acid

bicarbonate; hydrogen

chloride shift

in the arterial blood in the alveoli

bicarbonate; carbamino-hemoglobin

Conclusion

Congratulations! You have completed the program. If you have worked conscientiously, you should now be able to:

- recognize the elements present in various biological compounds.

- understand the term *pH* as applied to fluids and cells.

- recognize what is meant by the terms *acids*, *bases*, and *salts* as they occur in plant and animal tissues.

- know what electrolytes are so that their functions in the life of the cell can be more clearly understood.

- understand oxidation and reduction as they occur in the metabolic processes of plants and animals.

- differentiate among carbohydrates, lipids, and proteins, and recognize each from its unique functional groups.

- understand how enzymes function.

- know what nucleic acids are and how they are held together.

- compare diffusion and osmosis.

- distinguish between passive and active transport.

- understand osmotic pressure.

- understand transmission of nerve impulses, including depolarization and repolarization.

- understand how neurotransmitters work.

- know how DNA replicates.

ᴑ know how mRNA is formed and how it regulates protein synthesis.

ᴑ understand reactions involved in photosynthesis, including light reactions and the Calvin cycle.

ᴑ follow the flow of oxygen from the lungs to the cells and the flow of carbon dioxide from the cells to the lungs based on gas partial pressure gradients.

Appendix

Figure A
Cellular Respiration

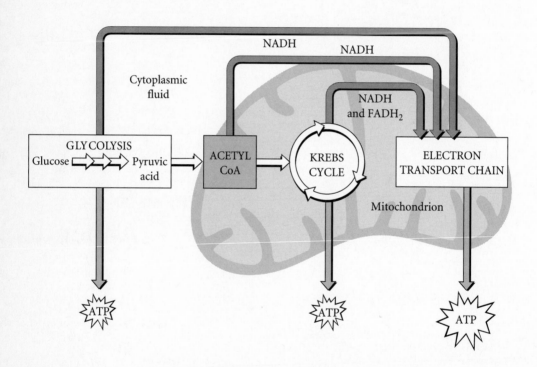

Figure B
Glycolysis

Glucose

ATP

Step **1**

ADP

Glucose 6-phosphate

2

Fructose 6-phosphate

ATP

3

ADP

Fructose 1,6-diphosphate

4

Glyceraldehyde 3-phosphate
(G3P)

2 NAD$^+$

5

2 NADH
+ 2 H$^+$

2 (P)

1,3-Diphosphoglyceric acid
(2 molecules)

2 ADP

2 ATP

6

3-Phosphoglyceric acid
(2 molecules)

7

2-Phosphoglyceric acid
(2 molecules)

2 H$_2$O

8

Phosphoenolpyruvic acid
(2 molecules)

2 ADP

2 ATP

9

Pyruvic acid
(2 molecules per
glucose molecule)

Figure C
Conversion of Pyruvic Acid to Acetyl Coenzyme A

Figure D
The Krebs Cycle

Figure E
Electron Transport Chain

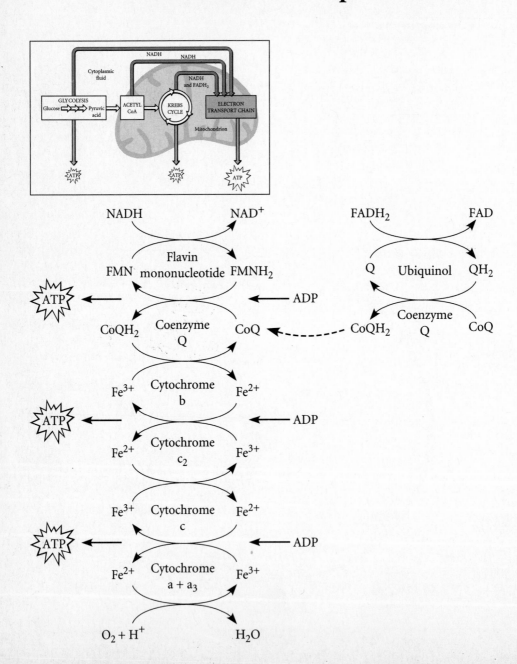